CONFLICT AND COOPERATION IN NATIONAL COMPETITION FOR HIGH-TECHNOLOGY INDUSTRY

A Cooperative Project of the

Hamburg Institute for Economic Research
HWWA

Kiel Institute for World Economics
IfW

and

National Research Council
NRC

on

"Sources of International Friction and Cooperation
in High-Technology Development and Trade"

NATIONAL ACADEMY PRESS
Washington, D.C.   1996

NATIONAL ACADEMY PRESS • 2101 Constitution Ave., N.W. • Washington, DC 20418

The National Academy of Sciences is a private, nonprofit society of distinguished scholars engaged in scientific and engineering research, dedicated to the furtherance of science and technology and to their use for the general welfare. Upon the authority of the charter granted to it by Congress in 1863, the Academy has a mandate that requires it to advise the federal government on science and technical matters. Dr. Bruce Alberts is president of the National Academy of Sciences.

The National Academy of Engineering was established in 1964, under the charter of the National Academy of Sciences, as a parallel organization of outstanding engineers. It is autonomous in its administration and in the selection of its members, sharing with the National Academy of Sciences the responsibility for advising the federal government. The National Academy of Engineering also sponsors engineering programs aimed at meeting national needs, encourages education and research, and recognizes the superior achievements of engineers. Dr. William A. Wulf is interim president of the National Academy of Engineering.

The Institute of Medicine was established in 1970 by the National Academy of Sciences to secure the services of eminent members of appropriate professions in the examination of policy matters pertaining to the health of the public. The Institute acts under the responsibility given to the National Academy of Sciences by its congressional charter to be an adviser to the federal government and, upon its own initiative, to identify issues of medical care, research, and education. Dr. Kenneth I. Shine is president of the Institute of Medicine.

The National Research Council was organized by the National Academy of Sciences in 1916 to associate the broad community of science and technology with the Academy's purposes of furthering knowledge and advising the federal government. Functioning in accordance with general policies determined by the Academy, the Council has become the principal operating agency of both the National Academy of Sciences and the National Academy of Engineering in providing services to the government, the public, and the scientific and engineering communities. The Council is administered jointly by both Academies and the Institute of Medicine. Dr. Bruce Alberts and Dr. William A. Wulf are chairman and vice chairman, respectively, of the National Research Council.

Limited copies are available from:
Board on Science, Technology,
  and Economic Policy
National Research Council
2101 Constitution Ave., N.W.
Washington, DC 20418
202-334-2200

Additional copies are available for sale from:
National Academy Press
Box 285
2101 Constitution Ave., N.W.
Washington, DC 20055
800-624-6242
202-334-3313 (in the Washington Metropolitan Area)

Library of Congress Catalog Card Number 96-70949
International Standard Book Number 0-309-05529-6

For the National Research Council, this project was overseen by the Board on Science, Technology and Economic Policy (STEP), a new standing Board of the NRC established by the National Academies of Sciences and Engineering and the Institute of Medicine in 1991. The mandate of the STEP Board is to integrate understanding of scientific, technological, and economic elements in the formulation of national policies to promote the economic well-being of the United States. A distinctive characteristic of STEP's approach is its frequent interactions with public and private sector decisionmakers. STEP bridges the disciplines of business management, engineering, economics, and the social sciences to bring diverse expertise to bear on pressing public policy questions. The members of the STEP Board are listed below:

NATIONAL RESEARCH COUNCIL
BOARD ON SCIENCE, TECHNOLOGY AND ECONOMIC POLICY

Sponsors

The National Research Council gratefully acknowledges
the support of the following sponsors:

The German-American Academic Council

Northern Telecom Limited

MEMC Electronic Materials, Inc.

Trimble Navigation

Samsung Electronics Co., Ltd.

Varian Associates, Inc.

Hitachi, Ltd.

Siemens Corporation

Philips Electronics N.V.

AT&T

General Electric Company

Program Support for the Board on Science, Technology and Economic
Policy is provided by a grant from the Alfred P. Sloan Foundation

Any opinions, findings, conclusions, or recommendations expressed in
this publication are those of the authors and do not necessarily reflect
the views of the project sponsors.

Table of Contents

v

I. Preface

PROJECT ORIGINS

This project has its roots in the belief that relationships need to be tended in order to remain strong and vibrant. Reflecting his concern with the vitality of the German-American intellectual dialogue, Chancellor Helmut Kohl proposed to then President George Bush that a sustained effort be undertaken to strengthen intellectual ties between the two countries. President Bill Clinton subsequently endorsed this concept and the German-American Academic Council (GAAC) was established in March 1993. The purpose of the Council is to support cooperation between Germany and the United States in all fields of science and scholarship by providing a forum for transatlantic dialogue and by collaborating on policy studies on issues confronting decisionmakers in both countries.

As its first policy project, the GAAC chose to sponsor an examination of the development of new technologies and the industries based on them. These technologies and industries are sources of economic growth and high-wage employment; competition for high-technology markets makes them also a source of growing international friction that could undermine both the multilateral trading system and the tradition of shared scientific and technological information. In recognition of the importance of understanding the roles of competition, conflict, and cooperation in high-technology development and trade, the German-American Academic Council solicited and approved a proposal from the U.S. National Research Council, through its Board on Science, Technology and Economic Policy (STEP), in partnership with two leading German research institutions, the Institute for Economic Research in Hamburg (HWWA) and the Institute of World Economics (IfW) in Kiel.

It was recognized from the outset that policy questions related to trade, investment, technology development, and cooperative activities are essentially global in nature. Furthermore, to ensure that the project yields practical policy recommendations for national governments, a sustained effort was made to bring a variety of perspectives to bear, not only scholarly analysis and technical expertise, but also business management and government policymaking experience. Accordingly, an innovative structure was adopted to secure the broadest participation with respect to project guidance, finance, conferences, and related activities.

MULTINATIONAL STEERING COMMITTEE

In order to provide leadership and direction for the project, a Steering Committee of distinguished academics, leading business executives, trade and technology policy practitioners, and other experts was assembled. The Committee includes members from Canada, Japan, Germany, and other European countries as well as the United States. The diverse national perspectives and training of this distinguished Committee brought a multidisciplinary and global perspective to the complex issues considered by the project. Different perspectives have a value in their own right but by no means assure consensus. The Steering Committee discussions involved a sustained effort to identify the limits of consensus on a broad range of analytically difficult and often contentious issues of great consequence for international cooperation in science, technology, and trade. The Findings and Recommendations reflect the consensus reached by the Steering Committee on these issues.

PRIVATE SECTOR PARTICIPATION

The generous GAAC grant covered the costs of participation for the German institutes and provided a foundation for the fundraising effort required of the National Research Council to meet its different budgetary requirements as a private independent institution. The challenge of securing adequate funding was also seen as an opportunity to secure broad private sector participation in the information-gathering phase of the project.

Validating the project's concept and the GAAC's interest, the National Research Council succeeded in assembling, over a relatively short period, a group of private sponsors that is very diverse in terms of nationality, sector of activity, and corporate size. These corporate contributors and participants include companies based in the United States, the Netherlands, Japan, Canada, Korea, and Germany, with operations across a broad range of high-technology sectors such as consumer electronics, semiconductors, computers, telecommunications, turbines, and materials. The participating companies include Northern Telecom, Siemens, Hitachi, Samsung, Philips, General Electric, MEMC, Trimble Navigation, Varian Associates, and AT&T.

The substantive and financial contributions of the project sponsors were essential to the success of this undertaking. Without their financial support, the NRC could not have carried out a project of this scope and intensity. Equally important, the active participation of senior industry representatives from these sponsors and a wide range of other companies, as well as of academic experts and senior policymakers, helped ensure that the presentations and discussions of the conferences accurately reflected the genuine opportunities for increased cooperation, the realities of global commercial competition for high-technology markets, the national stakes inherent in this competition, and the resulting policy challenges.

THREE CONFERENCES AND A SYMPOSIUM

The structure of the project reflected its international orientation, with each of the three participating institutions responsible for one of the principal conferences, to which the NRC added a timely symposium. The goal of these conferences was to assemble a body of analysis on the principal issues of the project. The conferences brought together an exceptionally knowledgeable group of academic experts, leading industrialists, and responsible officials of national governments and supranational institutions to participate in the presentations and debate. The sequence of the conferences was designed to progress from analysis of the underlying theoretical issues, through an examination of particular high-technology disputes and cases of international cooperation, to consideration of best practices and future policy options.

Each institution independently organized meetings reflecting its particular analytical strengths, policy interests, and traditions. The first conference, *The Economics of High Technology Competition and Cooperation in Global Markets,* was held at the HWWA in Hamburg, Germany, on 2–3 February 1995. The second conference, *The Sources of Friction and Cooperation in High-Technology Development and Trade,* was hosted by the National Academy of Sciences on 30–31 May 1995 in Washington, D.C. The third conference, *Towards a New Global Framework for High-Technology Competition and Cooperation,* took place at the Kiel Institute of World Economics on 30–31 August 1995 in Kiel, Germany.

In addition, the STEP Board of the NRC elected to hold a symposium,

International Access to National Technology Development Programs, on 19 January 1995. This symposium was designed to ensure a balanced analysis of the issue of foreign participation in national technology programs. Proposals to restrict participation in U.S. government-sponsored programs had generated intense debate in early 1995. The symposium enabled the project to address, in a timely and effective fashion, this controversial element of its agenda.

DIFFERENT TRADITIONS JOINED

The institutional partners in this collaborative effort brought different perspectives and traditions to this undertaking. The two participating German institutes form part of a system for providing high-quality economic analysis of public policy issues to the German government and public at large. For example, both institutes are well known for their expertise in global trade developments and international economic policy. Both institutes make important contributions to German policymaking. The National Research Council, as the operating arm of the National Academies of Sciences and Engineering, advises the U.S. government on a broad array of public policy questions. In addition to the different perspectives within the Steering Committee on the substantive issues reviewed by the project, the German institutes and the NRC have quite different procedures for taking positions on public issues.

The National Research Council is obliged by its procedures and traditions to submit its reports to independent review by experts who were not involved in the preparation of the report. The parent bodies of the NRC, the National Academies of Sciences and Engineering, consider this review to be an integral and constructive part of the study process, enabling committees to test their rationales, conclusions, and recommendations before a report is released to the public. Only after the authoring committee is independently judged to have been responsive to the reviewers' comments is its report released by the National Research Council. In this case, the Recommendations and Findings were agreed to by all members of the Steering Committee. Section IV of the report was prepared by the NRC staff, on the basis of contributions by Steering Committee members, the commissioned papers, and conference discussions. All sections of the Report were subject to the NRC review process, with the exception of Section V, which was prepared according to the procedures of the two German institutes.

The German institutes have a different tradition, reflecting their role in German public policymaking. They are able to release reports on their own responsibility, and the institute presidents are authorized to take policy positions on behalf of their institutions. It is for this reason that the joint document produced by the German institutes, though an integral part of the report, was not subject to independent review. This joint report does, however, represent the views of the institutes and was an integral part of the Steering Committee deliberations. It is therefore of great relevance to the final Recommendations and Findings of the Steering Committee.

The Steering Committee met on four occasions. Three of the meetings were held in conjunction with the conferences, in which Committee members were the principal participants. The final deliberative meeting, which took place at the NRC in Washington in December 1995, took into account the conference papers, presentations, and discussions and the analysis prepared by the three institutions. In the course of this final meeting, the Steering Committee agreed to a comprehensive and significant set of recommendations on a series of interrelated and highly complex issues. The Recommendations and Findings underscore the importance of the subject matter and address specific issues of technology and trade policy, government support of research and development, and policies affecting international cooperation. The recommendations also highlight the need for additional information and identify specific areas that would benefit from further analysis. In the rare instance where no agreement was possible, the Steering Committee acknowledges its inability to achieve consensus on a recommendation.

FINAL PRODUCTS

The results of the project are in four parts. The proceedings of the conferences and symposium are being published by the respective host institutions in three separate volumes. This volume contains the Findings and Recommendations of the Steering Committee, and revised versions of the two reports considered by the Committee at its final meeting, the first prepared by the NRC staff, the second jointly prepared by the HWWA and IfW staffs.

The process of arriving at the agreements which constitute this report was not always easy. In the end, it was successful, largely as a result of the cooperative spirit, dedication, and hard work of the participants. Indeed, in a sense, the success of the project is perhaps itself a model and a lesson for those who wish to strengthen international cooperation.

Erhard Kantzenbach   Alan Wm. Wolff
Project Co-Chairman   *Project Co-Chairman*

II. Introduction

THE CONTEXT FOR THIS REPORT

As the next millennium approaches, we are crossing the threshold into a new era of rapid technological change. Whatever prior experience in human history one chooses to look at—the introduction of widespread use of the horse, iron, steam generation in the industrial revolution, or the internal combustion engine—there has been no period like the present. Biotechnology is yielding genetically altered foods, medicines, and animals. Microelectronics has placed watches on our wrists that contain hundred-year calendars and cost less than the price of a hamburger; noninvasive surgery by laser takes place in minutes; phenomenal computing power has been placed in the hands of schoolchildren. Computers and fax machines enable homes to be offices. The communication of information is increasingly possible from any one human to any other, wherever located, with unparalleled ease. The diffusion of many types of information can be, and often will be, instantaneous.

National governments have a very long history of direct involvement in the development of high technology. Government has been a prominent and, at times, the main driver of technological advance. The technology of jet engines, telecommunications, computing, microelectronics, and in many cases biological advances, each had its roots in a national program. With the end of the Cold War, and the expansion of the market paradigm throughout much of the world economy, leading-edge technological developments in industry have shifted very heavily from public purposes toward satisfying private sector demands for commercial applications. The relative importance of the role of the military as a driver of technology has diminished significantly, though its needs remain of central policy concern.

In the last few decades, as democracy triumphed over communism, various forms of market-driven economies triumphed over those economies that relied on centralized state planning as the primary determinant of investment of human and capital resources. Reinforced by budget restraints, many governments are retreating from the scope of their involvement in what have increasingly become—in many parts of the world—commercial endeavors. For example, changing philosophy and disappointing performance push toward the dissolution of telecommunications monopolies. Heavy subsidization of industrial projects is being curbed in many developed countries, and gross distortions of the market are on the decline. Trade measures at the border have largely disappeared or are scheduled for sharp reductions.

Nevertheless, over the last several decades, intervention by governments in the promotion of technology has increased, accentuating the commercial competition among nations. Prior to this competition for high-technology production, trade friction had been common in agriculture, textiles, and steel, where heavy employment content gave a political stake to the sector of production. In the recent past and in the near future, technology-based industries are seen as involving the highest stakes in international competition for high-growth industries and the quality employment they provide and promise. Information-based activity is becoming a vastly increased portion of national economies. Change is very radical. Vulnerability to change, seen by some to be increased by the openness of international borders, is becoming a more prominent subject for policy debates, though less frequently for careful analysis.

Past experience is not a reliable predictor of the future, but it will inform judgments as policymakers consider the most appropriate international framework of rules and processes under which governments will interact, how they will deal with friction, and how they will consider cooperation in the development of high-technology goods and services and in the resulting competition and trade. In the last fifteen years, a number of the most contentious commercial disputes among nations have been in high-technology goods. Most prominent were the Airbus dispute between the European Community and the United States and the semiconductor dispute between Japan and the United States. In each case, the government role in high-technology industries was a major source of serious friction. In each case, industrial targeting by governments attempted to alter commercial competition. There is considerable evidence that this era is not wholly past. Government intervention and government toleration of private organization of the market are not likely to disappear. Government intervention by relatively new competitors, such as Korea, Taiwan, and China, is changing the global competitive environment, as these countries rely on a blend of government investment and private initiative to alter the terms of international competition in their favor.

Given the success many of these countries have experienced with effective state intervention, it is unlikely that the intellectual power of the free-market philosophy and the constraints of the purse will prove sufficient to avoid all areas of friction in the future. The efforts of governments to foster their companies' participation in high-technology industries are unlikely to disappear, and the new entrants may well make the competition in some sectors more intense. Governments will remain concerned with the relative standing of their industries because of concerns over economic advancement and employment. Technology-based industries and national power are closely related. Steel performed this role for nearly a hundred years. In the last decades of the twentieth century, this role is played by information technologies. The absence of government involvement in high-technology development, investment, and trade cannot, therefore, be presumed.

It is the purpose of this Report to identify the issues raised by competition in high-technology industries, to cite the underlying economic considerations and the facts involved in some prior areas of cooperation and friction, and to begin the process of making recommendations as to an international framework of principles and rules that governments should consider invoking to effect cures where problems are likely to persist. The emphasis is on current policies and practices, with reference to prior cases of cooperation and friction. Although there is a growing theoretical basis for governments' concern and for their support of high-technology industries, the analysis here places less emphasis on underlying economic theory and more on the practical steps that policymakers might take to achieve legitimate societal goals while minimizing friction and engaging in cooperation with their rivals, partners, and neighbors.

There are growing pressures for greater cooperation—cost, technique, technology, market access, and shared risk drive cooperation across national borders. Exploiting these opportunities fully will require new norms. Wherever possible, the Report attempts also to assess the positive role of government in the development of high technology and in exploring opportunities for greater cooperation and understanding and for avoiding friction.

THE STEERING COMMITTEE'S
RECOMMENDATIONS AND FINDINGS

The Steering Committee of the Project, coming from many of the major nations involved in both the development of high-technology goods and services and trade in these goods and services, and having widely varying backgrounds in policy making, industry, research, and academic pursuits, concluded that friction and cooperation in high-technology development and trade will have important consequences for the community of nations, and deserve special attention. Action and inaction by policymakers will have important consequences for our citizens and for the global economy. There will be continuing competition for new technologies and new industry. In the appropriate international framework, global benefits can be maximized and friction reduced.

The Steering Committee's recommendations and findings pose a challenge to national governments and to persons in industry and academia having an interest in fostering the positive evolution of a multilateral trading system. Specifically, the framework should result in

- Markets for high-technology goods that are open and contestable through both trade and investment;
- Enhanced competition and cooperation across national borders;
- Government-supported research and development (R&D) for essential government functions that avoids distortion of markets;
- The formation of national and international consortia to reduce risks and costs associated with new technologies and standards;
- Increased openness of national programs to qualified foreign entities;
- Effective protection of international property rights to encourage innovation and commerce;
- The curbing of injurious subsidization;
- Elimination of distortions of international investment flows due to restrictions or excessive incentives;
- Elimination of other distortive government measures, such as tariffs, discriminatory public procurement, offsets, and exclusionary standards and certification requirements; and
- Prevention of the frustration, through private anticompetitive practices, of efforts to attain the above objectives.

The Steering Committee also called for a series of specific areas for further research to provide a sound basis on which policy could be made.

The members of the Steering Committee shared a sense that through the development of technology, the peoples of the world stand on a threshold of unparalleled promise and opportunities. Most gatherings of those interested in technology focus on the challenges and benefits of fostering a particular technology. This project differed in that it directed its attention to attempting to learn from past examples of both international cooperation and frictions, and to creating a global public policy environment that will make more likely the realization of benefits from technology through international cooperation and competition.

Alan Wm. Wolff
Project Co-Chairman

III. Steering Committee Policy Recommendations and Findings*

1. High-technology products and their development require particular attention in light of their importance to the national economy and because these industries are the target of industrial policies of many participants in the multilateral trading system.

OPEN MARKETS

2. Markets for high-technology industries should be open and contestable. Market openness through both trade and investment is a powerful means for improving national and global welfare.

3. Governments should seek to enhance competition, and thereby increase global efficiency, rather than resisting market forces and discouraging competition.

4. Public policymakers should avoid restricting, wherever possible, transnational cooperative efforts among firms, provided that sufficient competition is preserved.

---

*Recommendations and Findings Endorsed by all Members of the Steering Committee.

GOVERNMENT SUPPORT FOR RESEARCH

5.   Government support for public research infrastructure and its linkages to private sector research is essential for scientific and technological progress.   In addition, governments should support research and development for essential government functions.

6.   In general, social benefits of R&D exceed private benefits of R&D, providing justification for government support.

7.   When supporting R&D, governments should generally support R&D itself, and not target products or trade to avoid distorting markets.

8.   Government can serve as a facilitating agent to create the necessary credibility, commitment and mutual trust among private firms in the formation of research consortia.   It would be valuable for policymakers to learn from private sector experiments in structuring, and participating in, international consortia undertaken to reduce the risks and costs associated with the development of new technologies and standards for their application.

9.   Governments should be encouraged to allow qualified foreign entities' participation in national R&D programs which receive government support, on conditions which are transparent, mutually agreed, and mutually beneficial.   To the extent that international agreements have not been reached, governments' national benefits requirements for participants in national technology development programs should seek to avoid imposing constraints on private participants concerning the transfer or deployment of technologies resulting from such programs.

10. Effective protection of intellectual property is also a key element of viable programs of international cooperation in research and development.

TRADE ISSUES

11. With respect to the Subsidies Code of the General Agreement on Tariffs and Trade (GATT), governments should consider, in light of current experience, eliminating the special treatment (greenlighting)* now provided for R&D and certain other subsidies.

12. Effective and adequate protection of intellectual property rights is essential for encouraging innovation, and its commercial exploitation, throughout the global economy.

13. Governments must also address challenges to the existing intellectual property regime. These challenges include: differences in intellectual property regimes among major trading countries; effective enforcement in the rapidly industrializing countries; the issues posed by rapid technological change in high-technology industries such as biotechnology (e.g., the Human Genome

---

*Editor's note: "greenlighting" refers to subsidies that are not actionable under trade remedy provisions because their purpose is internationally accepted.

Project); and the global information infrastructure (e.g., copyright of digital information). International negotiations are needed to adjust existing norms.

14. Foreign direct investment remains a central issue in high-technology development and trade. It is a major channel for both technology and trade flows. An effective international investment regime based on a multilateral investment accord should promptly be completed. Excessive incentives in competition for new investments should be limited at the national, regional, and perhaps international level. Compulsory technology transfer must be prohibited. Market access through investment must be open.

15. Some policies designed to support high-technology industries are harmful to the multilateral trading system. In addition to investment restrictions and performance requirements, activities such as discriminatory public procurement, offsets, and counter trade, and exclusionary product standards and certification requirements should therefore be brought under international review and scrutiny.

16. Discrimination based on national origin in the public procurement of high-technology products and services should be minimized. To the extent it exists, it should be expressed as a clear price or cost margin differential, the levels of which should be bound* and then reduced or eliminated.

17. Standards competition has become increasingly important in high-technology sectors. In this context, new issues of market power and collusive behavior (including "fair" access for competing firms) may arise, posing problems for international competition policy. Standards-setting should not be designed to achieve anticompetitive results. In addition, government regulation and participation should not unduly delay the establishment of new standards.

18. For certain technology products, substantial tariff barriers still persist. A concerted multilateral effort should be undertaken to eliminate remaining tariffs in high-technology sectors by the year 2000, or sooner, and should be promptly agreed.

19. Existing barriers to market access for information technologies and the development of a global information infrastructure, e.g., restrictions on telecommunications investment and services, should be removed.

20. Structural impediments to trade must be eliminated. Existing GATT articles offer a multilateral route to conflict resolution. The nonviolation clause of Article 23 of GATT provides access to multilateral dispute settlement even when the defending country has not explicitly violated the GATT. Where actionable under the World Trade Organization (WTO), multilateral dispute settlement should be utilized. Where the WTO rules are inadequate, they should be amended to achieve these results.

21. No consensus could be achieved over dumping and antidumping issues.

---

*Editor's note: contractually committed to by international agreement.

COMPETITION POLICY

22. Anticompetitive business practices, both within national economies and in the international system, can have severe trade-distorting effects. A sustained effort to develop principles and practical guidelines for international cooperation with respect to competition policy should be undertaken as a natural extension of international trade principles.

23. In this regard, three steps should be taken: Minimum standards should be developed, to be observed by antitrust authorities in the participating countries in dealing with private restraints of international trade. Multilateral means should be developed to ensure the application of national law by an inactive member state where it will not act on its own initiative. Governments should consider applying WTO dispute settlement to anticompetitive activities which, occurring in one signatory country, have adverse effects in another.

FURTHER WORK

24. To address the issues raised by international competition among national economies for high-technology industry, governments must ensure adequate information and analysis on which to base policy. A number of issues requiring further study were raised in the course of our discussions, including

• Further analytical work concerning the principles of effective cooperation in technology development, the lessons from national and international consortia, including eligibility standards, should be undertaken. This work could also include assessments of what new cooperative mechanisms might be applied to meet the challenges of international cooperation in high-technology products.

• Further work on the role and support for the university and research institutes in international science and technology research should be undertaken.

• Further work on current trends in public and private support for basic research is recommended.

Specific projects related to research and development and trade in high-technology goods and services might include further analytical work in

• Government procurement
• Intellectual property protection
• R&D subsidies
• Standards competition
• Investment restrictions and performance requirements
• Private restraints of trade

IV. NRC Summary Report on the Project

Foreword to the NRC Summary Report

This section (IV) and the following section (V) represent contributions prepared, respectively, under the auspices of the NRC and the HWWA and IfW. Each reflects a synthesis of the issues and considerations reflected in the Joint Recommendations and Findings, drawing upon the conferences, symposium, and the deliberations of the Steering Committee. These sections address the main issues raised in the course of the project on **International Friction and Cooperation in High-Technology Development, Competition, and Trade.**

The analysis presented in this section (IV) is specifically drafted with the policymaker in mind. Its focus is on improving international understanding of the sources of conflict and cooperation in markets for high-technology goods and services. It describes the motives for national efforts to develop and support high-technology industry in order to capture the economic and political benefits of such activity for national economies. This section then examines the motives for greater national and international cooperation in the development of new technologies and reviews the challenges such cooperation can pose to participants and policymakers alike. Important new trends such as strategic alliances, cooperation between national development programs, and the globalization of R&D are also reviewed.

Taking a global perspective, this section reviews both the interaction between national efforts to support high-technology industry and the impact of these practices on the multilateral trading system. It discusses the importance of practices such as discriminatory public procurement, dumping, intellectual property protection, the growth in investment incentives, and the importance of differences in national investment regimes. Current policy approaches, both bilateral and multilateral, are discussed. These include the importance of competition (or antitrust) policy, the need for a multilateral investment accord, and more generally the potential role of national policies in advancing multilateral solutions. Examples of government policies and practices and private sector strategies are outlined as a means both of illustrating the real-world relevance of the issues under discussion, and of pointing to potential lessons for future policymaking.

The analysis both supports the Recommendations and Findings and is intended to contribute to the debate over how to maintain the multilateral system in the context of increased national competition for high-technology industries. Both this section of the Report and section V have benefited immensely from the willingness of Steering Committee members to make available their time, expertise, and experience.

I would like to particularly thank my colleagues on the Steering Committee who made important contributions to this project. A number of these individuals can fairly be associated with the analysis set forth in this section although some may have reservations on particular points. This section benefited greatly from the spirited discussion among all the participants, and particularly our HWWA and IfW colleagues. The participants would certainly not agree on all points, nor in the matter in which they should be expressed, but I believe that our common objectives, our joint Recommendations and Findings, formally set forth in the preceding section, far outweighed any differences. The latter were always offered with zest, and only served to sharpen our thinking on the more difficult issues.

Special recognition for this section and for American participation in this project is due to Dr. Charles W. Wessner, the principal author of this analytical section of our Report. In addition to his written contributions, he brought his considerable policy expertise, unflagging energy, and remarkable organizational skills to bear on this complex international undertaking. Without his efforts, this project would not have been undertaken or successfully concluded. Invaluable assistance in the preparation of the NRC conferences and the drafting of the main body of the Report was provided by George Georgountzos, the program associate for the Board on Science, Technology, and Economic Policy (STEP). Their extraordinary dedication and commitment were essential to the successful completion of this project.

Other members of the STEP team, to whom our appreciation is owed, include Dr. Stephen A. Merrill, who played an instrumental role in the initial phases of the project, and Lena Steele, who made important contributions to the May 1995 NRC conference. The experience and timely advice of Dr. Lawrence E. McCray were welcome contributions at key points in the project. Recognition is also due to Patrick Stuart and Anne Eisele, both of whom provided vital support in the initial phases of the project and especially for the January 1995 symposium. Last, but not least, the optimism, support, and creativity of Dr. E. William Colglazier played a crucial role in bringing this project to a successful conclusion.

Alan Wm. Wolff
Project Co-Chairman

Sources of Friction and Cooperation in High-Technology Industries

During this century, scientific discovery and engineering developments have brought enormous technological progress, with widespread benefits for mankind in both the industrialized and industrializing world. As a recent report of the National Research Council remarked, "the change has often been gradual, almost unnoticed in daily life, but fundamentally important."[1] Technological advance has permitted astronauts to walk on the moon and astrophysicists to probe the origins of the universe, while the physical sciences have brought us microelectronic devices, lasers, and fiber-optic networks. High-technology industries such as aircraft, chemicals, computers, software, pharmaceuticals, and biotechnology continue to bring us new products that lengthen our lives, extend our personal ability to communicate and access knowledge, and increasingly to provide cures to the ills that plague man.

The power—for it is that—and wealth-creating activities generated by these technologies also bring out the acquisitive instincts of mankind, engendering competition which holds the twin potential of greater excellence and debilitating conflict. This Report reviews some of the sources of international competition for high-technology industry, the problems and risks such competition entails for scientific and economic relations, especially trade, the forces encouraging greater international cooperation, and the challenges these cooperative efforts encounter.

THE PERMANENCY OF COMPETITION FOR HIGH-TECHNOLOGY INDUSTRY

Competition over strategic high-technology industry will continue to be a major source of friction in the international system.[2] While a degree of healthy competition is inevitable and desirable, unless sustained and effective attention is given to the complex set of policy issues associated with the programs to develop and nurture high-technology industry within national economies, the friction generated by these competing national programs could have important

---

[1] *Allocating Federal Funds for Science and Technology,* National Research Council, National Academy Press, Washington, D.C., 1995, p. 70.

[2] It is important to note at the outset that the composition of world trade has changed dramatically in the twentieth century. In 1900, world trade was approximately 80 percent agricultural and mineral based; today it is 80 percent based on manufactures and services. This means that the advantage on which specialization and trade is based are roughly 80 percent man-made; trade based on natural resource endowments is of little significance (see Figure 1). High-technology trade is an extreme example of this general development, with government initiatives playing a major role.

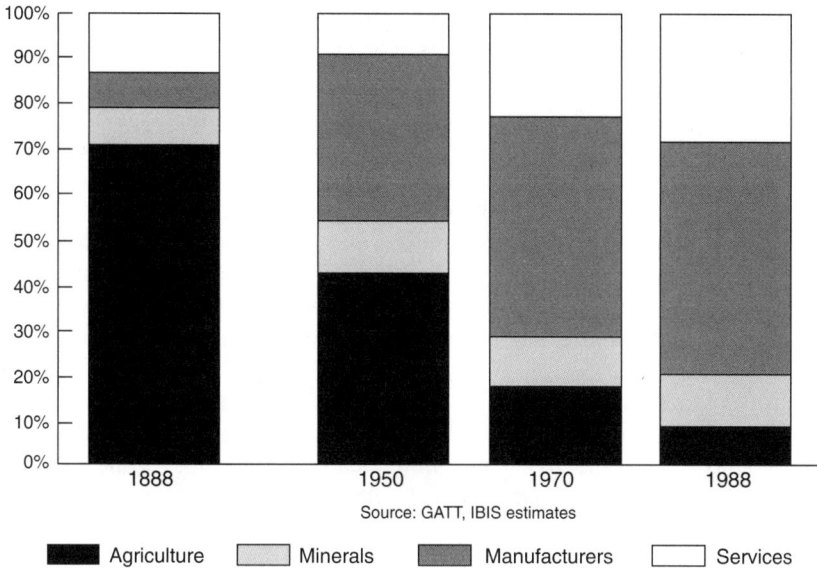

FIGURE 1   World trade in goods and services. The chart shows shares held by main commodity groups. From B. Scott, "Economic Strategy of Nations."

negative consequences for scientific and technological cooperation and for the international trade regime. It is important to note at the outset that the United States, Europe, Japan, Korea, Taiwan, China, and other major trading countries all have extensive national programs to support basic and applied research. These programs frequently include the development of core technologies in sectors such as electronics, information systems, aerospace, new materials, and opto-electronics, sectors that have a decided impact on commerce. While national and regional programs vary greatly by structure, funding, participation, and rationale, their basic objectives are similar.

GROWTH IN REGIONAL AND NATIONAL
TECHNOLOGY DEVELOPMENT PROGRAMS

The European Community is increasingly concerned about the gap between its acknowledged scientific excellence and its ability to translate this asset into practical economic and commercial achievement. The relative decline in the position of European industry in technology-intensive sectors such as microelectronics and computers and the limited commercialization of biotechnology are of concern to European officials at both the national and regional levels. A recent European Commission report observes that Europe's capacity for innovation has diminished in recent decades. It notes with "alarm" that, "between 1981 and 1993, the number of patents filed in Europe steadily declined in several key sectors: electronics, pharmaceuticals, chemicals, the aircraft industry, etc. For Japan in the same period, the number of patents showed a steady increase." The report also notes that "advanced technology products only account for about 30% of the Union's exports, while the corresponding rate for Japan and the United States is over 50%."[3] The report cites three features of the European research system to partly explain these weaknesses: the inadequate translation of research results into commercial applications, insufficient investment in research and technology development programs in the fields of education and training, and the fragmentation and lack of coordination in European research efforts.[4]

European regional efforts to redress their competitive position include well-known endeavors such as the European Union programs in information technologies (ESPRIT); advanced communications technologies and services (ACTS); industrial and materials technologies (BRITE); standards, measurement, and testing (SMT); and the marine science and technology program (MAST), as well as transport research (for air, rail, road, and integrated transport). These programs, which represent a significant commitment of Community funds, are focused on the development of precommercial and commercial technologies—not basic research.[5] These programs were supplemented in the mid-1980s by the European Research Technology Program (EUREKA). Conducted outside of the European Community institutions, the program consists of joint R&D projects in advanced technologies involving more than one member state, such as the Joint European Submicron

---

[3] European Commission, *Research and Technology: the Fourth Framework Programme (1994–1998),* Brussels, Belgium, 1995, p. 12.

[4] Ibid. The lack of coordination and the risk of needless duplication are a recurrent—and understandable—theme of the European Commission. While no doubt some wasteful duplication occurs at the national level, given the uncertainties associated with technology development and the innovation process in general, it is not certain that a centralized approach along the lines of the European Framework model is inherently superior—hence the existence of the EUREKA program as well as major programs at the national level (see below). For a review of European Union programs, see W. Sandholz, *High-Tech Europe: The Politics of International Cooperation,* University of California Press, Los Angeles, 1992.

[5] The European Community programs, such as the Framework programs, are focused not on basic science but on the improvement of industrial competitiveness. This orientation is based on article 130F of the Treaty on European Union, which notes, "the Community's aim shall be to strengthen the scientific and technological basis of European industry and to encourage it to become more competitive at the international level." European Commission, *The European Report for Science and Technology Indicators, 1994,* Directorate-General XIII, Telecommunications, Information Market and Exploitation of Research, Luxembourg, October 1994, p. 257.

Silicon Initiative (JESSI). Despite the prominence given to Europe-wide programs, many of the most important policies and programs in terms of resources, focus, and competitive impact remain at the national level.[6]

The United States is undergoing a structural transformation in its national research programs largely as a result of the end of the Cold War. The structural change engendered by this historic transition has been compounded by the more rapid technological advance of the commercial sector—a reversal of the leading role previously held by the military production base. (See Supplement C on dual use technology below.[7]) This evolution in U.S. technology development programs has by no means eliminated the government role in technology development. Indeed, dual-use strategies emphasize the importance of the commercial-industrial base in the effort to provide the most advanced technologies on a timely, cost-effective basis. Consequently, the military rationale for U.S. government support for technology—which served to justify government investment throughout the Cold War—is now joined by the growing awareness of the competitive pressures the U.S. economy faces as well as by the dramatic increase in the cost of developing new technologies.[8]

As a result, the U.S. has a broad and growing portfolio of cooperative initiatives such as the dual-use oriented Technology Reinvestment Project (TRP, which includes programs such as the U.S. Display Consortium), the

---

[6] For a review of the different perspectives on and evolution of national policy concerning "state of the art" technology within Germany, as well as a discussion of the different views on such programs within the EU, see Erhard Kantzenbach and Marisa Pfister, "National Approaches to Technology Policy in a Globalizing World Economy: The Case of Germany and the European Union" in G. Koopman and H.E. Scharrer (eds.), *The Economics of High-Technology Competition and Cooperation in Global Markets,* HWWA Institute for Economic Research, Hamburg, Germany, 1996. For a comprehensive overview of European science and technology programs, see *The European Report on Science and Technology Indicators 1994.* For a review of programs at the national level in semiconductors, see Thomas Howell, Brent Bartlett, and Warren Davis, *Creating Advantage: Semiconductors and Government Industrial Policy in the 1990s,* SIA, Santa Clara, Calif., 1992.

[7] The term dual-use refers to research and technologies that have both military and civilian applications. In addition to the dual-use section below, see the supplement on the Global Positioning System.

[8] A recent report by the National Science Foundation notes that U.S. trade in several advanced technologies, including aerospace, computer-integrated manufacturing, life science, and computer software, produced sizable trade surpluses in the 1990s—*yet this surplus has declined every year since 1991* (italics added). The same report notes that "U.S. technology trade is highly concentrated," with 85 percent of total U.S. technology product exports in information technologies, aerospace, and electronics. National Science Board, *Science and Engineering Indicators—1996,* U.S. Government Printing Office, Washington, D.C., 1996, NSF PR 96-22, 20 May 1996, p. 6-2. The report adds that recent shifts in industrial research and development in the United States and abroad are narrowing the margin of technological advantage for U.S. firms.

SEMATECH consortium, the High-Performance Computer and Communications (HPCC) program, and the multifaceted National Information Infrastructure Initiative (NII), the Small Business Innovation Research Program (SBIR), the Advanced Technology Program (ATP), and the more recent Partnership for the New Generation of Vehicles (PNGV) program.[9]   In addition, there has been rapid expansion of cooperative programs, such as the Department of Energy's (DOE) Cooperative Research and Development Agreements (CRADAs) for technology cooperation between DOE laboratories and private firms. Many of these programs involve government-industry partnerships with shared costs, management, and objectives to develop technologies to meet both government missions and commercial goals.[10]

At the same time, there are contradictory trends in the funding of the U.S. R&D effort. Notwithstanding the relatively broad consensus concerning the value of government support for research, the constrained budgetary situation in the United States has resulted in strong pressures for the reduction of important elements of the U.S. federal R&D portfolio, with the multiyear budgets of both parties implying substantial redutions in funding for research and development.[11]   The debate in the United States concerns not only the *level* of support for R&D, but also the *composition* of government support. Some influential U.S. policymakers oppose government support for technology development programs, arguing that there should be a clear line between basic and generic research (which all agree government should support) and applied research.[12]   Others argue that the research process

---

[9] For a review of the PNGV program, see *Review of the Research Program of the Partnership for a New Generation of Vehicles (PNGV)*, National Academy Press, Washington, D.C., 1994. For the NII, see *The Unpredictable Certainty: Information Infrastructure Through 2000*, National Academy Press, Washington, D.C., 1996. The much-discussed U.S. Advanced Technology Program (ATP) began under the Bush administration, was increased rapidly under the Clinton administration, and subsequently encountered substantial criticism. However, this program remains relatively small, both in absolute amounts ($340.5 million in 1995) and in terms of the size of the awards. For a discussion of the ATP program in the context of U.S. technology policy, see Robert M. White, *U.S. Technology Policy: the Federal Government's Role*, Competitiveness Policy Council, Washington, D.C., September 1995.

[10] For a review of U.S. partnership programs, including recommendations for improving their effectiveness, see Richard J. Brody, *Effective Partnering: A Report to Congress on Federal Technology Programs*, Office of Technology Policy, U.S. Department of Commerce, April 1996.

[11] The U.S. Council of Economic Advisers argues, however, that investments in R&D are the key to increasing productivity. A recent CEA report notes that "successful R&D investments—from the jet engine to transistors to lasers—can and have transformed the whole economy." See *Supporting Research and Development to Promote Economic Growth: The Federal Government's Role*, Council of Economic Advisers, Washington, D.C., October 1995, p. 8 and *passim*.

[12] Ibid.

does not fit into neat categories.[13] They argue that, in reality, the process of invention and application is a continuum, with many applied research projects yielding significant returns to society at large.[14] Whatever the merits of the debate, the paradox remains that while U.S. government support for technology development appears to be in an innovative phase, the absolute amount of funding for U.S. research and development is declining.[15]

Japanese policy has had a long-standing national orientation toward acquiring, diffusing, and refining new technologies. Japanese industrialists and policymakers alike recognize the importance of new technologies for economic growth and national competitiveness and have developed a wide variety of supportive policies. For example, the Ministry of International Trade and Industry (MITI) and its implementing agencies have carried out major projects in areas such as Supersonic Jet Propulsion, Very Large Scale Integrated Circuits for semiconductors and related materials, and the Fifth Generation Computer Project. They also launched the initial Intelligent Manufacturing Systems Partnership, which has developed into a substantial international cooperative effort.[16]

---

[13] See *Allocating Federal Funds for Science and Technology,* especially Supplement 4, pp. 70–81. The report observes that major firms such as Sun Microsystems, Silicon Graphics, Genentech, and Amgen "did not exist fifteen years ago. All were started from a base of academic science." (p. 77) The report cites the Nobel laureate Lord Porter, who observed, "there are two kinds of research—applied research and not-yet-applied research." See also the discussion of subsidies below.

[14] See Box A. Economists argue that R&D has high private rates of return and even higher social rates of return, that is, "benefits which accrue as other researchers make use of new findings, often in applications far beyond what the original researcher imagined." Council of Economic Advisers, *Supporting Research and Development to Promote Economic Growth,* p. 5.

[15] U.S. industrial investment in R&D fell in the first half of the 1990s, at the rate of about 1.5 percent a year in constant dollars. NSF PR 96-22, 20 May 1996. See National Science Board, *Science and Engineering Indicators 1996.* This trend in the U.S. R&D effort is a cause for concern at the upper levels of the U.S. government. See, for example, the statement by Anita Jones, director of Defense Research and Engineering, U.S. Department of Defense to the conference *Sources of International Friction and Cooperation in High-Technology Development and Trade, 30–31 May 1995.* National Academy Press, Washington, D.C., forthcoming.

[16] For a review of Japanese programs and policies, particularly with respect to the efforts for greater international cooperation that characterized the early 1990s, see Gregory Rutchik, *Japanese Research Projects and Intellectual Property Laws,* Office of Technology Policy, U.S. Department of Commerce, Washington, D.C., 1995. For a discussion of the IMS partnership, see the presentations by Robert Cattoi and Uzuhiko Uwatoko at the conference *Sources of International Friction and Cooperation in High-Technology Development and Trade.* Robert Cattoi described the IMS as "a catalytic agent for global manufacturing cooperation involving large and small companies, user and suppliers, universities and governments." See *Intelligent Manufacturing Systems,* Coalition for Intelligent Manufacturing Systems and the U.S. Department of Commerce Technology Administration, Washington, D.C., 1995.

Although Japan's technology programs are seen as highly successful by many European and American policymakers, reflecting Japan's rapid movement to the forefront of many advanced technologies, Japanese policymakers perceive their economy as facing major challenges both from industrial countries, such as the U.S., and newly industrializing countries, such as Korea.[17] Reflecting this perception, Japan has recently announced—in sharp contrast to trends in the United States—plans to double its R&D spending by the year 2000.[18] For example, 1997 funding for industrial research is to increase more rapidly than the overall budget, with MITI's Agency for Industrial Science and Technology (AIST) receiving a significant increase, reflected by AIST's announcement of three new programs.[19] Because of the central role of semiconductors, as the enabling technology for the microelectronics revolution, the long-standing commitment of the Japanese government to this sector is expanding.[20] For example, Japan recently

---

[17] For a Japanese view of the problems faced by their industry, see M. Sumita and H. Shin, "Japan's Semiconductor Industry in the 21st Century," *NRI Quarterly,* Spring 1996. They observe that "semiconductor demand is booming worldwide, but Japan's semiconductor industry is in trouble—squeezed by a resurging American chip industry and the growing inroads of Korean companies." They believe that "the American semiconductor industry has successfully differentiated its products from those of its competitors, including Japan..." They also affirm that "much of its success can be traced, in particular, to (1) joint projects with the leading users of semiconductors to develop new products; and (2) close cooperation between the public and private sectors in support of the semiconductor industry. The Korean industry has focused on developing and producing commodity-type semiconductors...and the Koreans now have a significant cost advantage over the Japanese." Ibid., p. 20.

[18] The contribution of Japanese industry to the national R&D effort should not be understated. In the 1970s, Japanese electronics firms were still acquiring basic technologies but rapidly developed new ways to apply them. By the 1980s and 1990s, the larger firms actively carried out their own basic research while remaining at the forefront of applications. Recognizing the importance of creating core technologies, companies such as Hitachi regularly invest 10 percent of sales in R&D activities while engaging in a broad range of international alliances. See the presentation by Y. Takeda, "Japanese Technology Acquisition, Diffusion, and Development Firm Strategy: Changes and Challenges" to the conference *Sources of International Friction and Cooperation in High-Technology Development and Trade, 30–31 May 1995.*

[19] These include (1) an industrial technology and innovation R&D program ($25.2 million); (2) the "dream project," which will include a structural biology component and research on next-generation optical memory technology ($1.5 million); and (3) the "techo-infra" project to establish measuring and testing standards and to create an information infrastructure for biological resources ($9.3 million). See U.S. Embassy, Tokyo, 960306, "Research Budget," in *International Market Insight Series.*

[20] Japan is by no means the only country to share this policy focus on the semiconductor industry. There are now more than ninety semiconductor industry research organizations worldwide. Current estimates include twenty-five in Europe, eighteen in Japan, seventeen in Korea, fifteen in Taiwan, and some ten in the United States. Currently, China has only two, though this number is to increase. Excluding the United States, total five-year funding for electronics research now approximates $80 billion. (See below.)

announced a series of new consortia for semiconductor research.[21] Given the competitive pressures faced by Japanese high-technology industry, Japan's positive experience with government-industry programs, and the perception that collective efforts of other nations have proved successful, new and expanded Japanese programs to meet the technical and capital challenges of new technologies can be expected.

GREATER NATIONAL COMPETITION

This overview suggests that national governments and regional authorities are likely to continue and intensify efforts to locate high-technology industry within national borders (see Table 1). The widespread conviction that these industries offer the greatest prospect of substantial economic growth, including high wage, high value-added employment, continued technological competency, and enhanced national autonomy, is unlikely to diminish.[22] Among the industrial countries, concerns about structural unemployment, particularly in Europe, combined with uneasiness about the employment consequences of increasingly streamlined industries, and the rise of high-quality, low-cost foreign competitors (in both traditional and "new" high-technology industries) generate powerful political pressures which have led governments to adopt more activist policies to nurture and protect national industries, especially in strategic sectors such as aerospace and electronics.[23] This has added a new dimension to the traditional competition between Europe, the United States, and Japan with regard to certain high-technology industries.

---

[21] They are, inter alia, the program for Semiconductor Leading Edge Technologies (SELETE), the Semiconductor Technology Academic Research Centers (STARC), and the Super Silicon Crystal Research Institute (SiSi). The SELETE program is a Japanese-only effort to develop the equipment and materials for the move to the 300mm semiconductor wafer standard. Presentation by Hiroyoshi Komiya, executive vice president and chief operating officer, SELETE, at the conference "The U.S., Japan, and the Rules of the Game in High-Technology: National Policies and International Competition in Semiconductors," the Brookings Institution in cooperation with Nomura Institute, 9 May 1996. See the section on International Cooperation below and Box D on International Cooperation on the 300mm wafer. Interestingly, these new Japanese programs appear to be modeled on SEMATECH, though with substantially greater funding.

[22] Alan Wm. Wolff, Thomas R. Howell, Brent L. Bartlett, and R. Michael Gadbaw (eds.), *Conflict among Nations: Trade Policies in the 1990s,* Westview Press, San Francisco, 1992, p. 528.

[23] Sylvia Ostry and Richard Nelson, *Techno-Nationalism and Techno-Globalism: Conflict and Cooperation,* The Brookings Institution, Washington, D.C., 1994, pp. 60–78. In addition, the authors observe that, paradoxically, government measures to encourage national industry may also spur greater alliance activity and increased local investment by foreign competitors.

TABLE 1. Major Research Consortia in Japan, the United States, and Europe*

Name	Country	Dates	Budget ($m)[a]	Government share (%)
VOLS.	Japan	1976-80	350	40
OMCS	Japan	1979-85	90	100
VHSIC	USA	1980-89	900	100
Supercomp.	Japan	1981-89	130	100
FED	Japan	1981-90	40	100
SG	Japan	1982-91	426	100
Alvey	UK	1983-88	500	50
ESPRIT I	Europe	1984-89	1800	50
ESPRIT II	Europe	1988-93	3800	50
Eureka	Europe	1985-96	7700	50
RACE	Europe	1985-96	3000	50
JESSI	Europe	1989-96	4000	50
MCC	USA	1983-	80 [b]	0
NOMS	USA	1986-	150 [b]	50
SEMATECH I	USA	1987-92	1000	50
SEMATECH II	USA	1993-98	200 [c]	50
ASET [d]	Japan	1996-2001	100 [b]	100
SELETE	Japan	1996	350	50
STARC	Japan	1996-2000	10	n/a
SiSi.[e]	Japan	1996	70	50
I300I [f]	International	1996-?	40 (est.)	0

[a] Total amounts (government plus industry)

[b] per annum

[c] The 1995 government contribution was $85 million. At the request of its Board, SEMATECH will no longer receive government funding after 1996.

[d] Japanese subsidiaries of 3 U.S. firms are among 21 corporate participants.

[e] Super Silicon Crystal Research Institute to develop 400mm wafers.

[f] Current participants in this 300mm wafer project include companies from Korea, Taiwan, Europe, and the U.S.A. SELETE is the parallel Japanese national 300mm wafer program. See Box C.

* Drawn from Peter Grindley, David C. Mowery, and Brian Silverman, "SEMATECH and Collaborative Research: Lessons in the Design of High-Technology Consortia," *Journal of Policy Analysis and Management, vol. 13, no. 4,* 1994, p. 727, with supplemental information from the NRC and Kenneth Flamm, *Mismanaged Trade: Strategic Policy in the Semiconductor Industry,* The Brookings Institution, Washington, D.C., 1996, pp. 437–441.

This competition is increasingly vigorous, and has the potential to place strains on otherwise satisfactory political and economic relationships. However vigorous, it is nonetheless not a military competition, but rather "a peaceful competition among those who were fated by their natures to be economic rivals... More is at stake in this competition than employment; were that the sole issue, macroeconomic means might well accomplish the desired results. The large industrial efforts of the...major trading nations are not evenly spread among the various product sectors for purposes of job creation. They are targeted at certain sectors viewed as strategic. Governments believe that the future of their countries depends on the composition of their economies, and for the most part they see their success as nations defined by their relative success in these specific efforts."[24]

In addition, the high-technology competition among the established industrial powers is being profoundly modified by the emergence of new entrants wishing to compete for the high-technology industries which were previously reserved to the most advanced countries. These new entrants are altering the terms of global economic competition with policies different in important ways from the practices and proscriptions of the leading countries. New state-supported producers in Korea, Taiwan, Malaysia, and, increasingly, China are aggressively entering global markets for high-technology products.[25] India is also rapidly emerging as a participant in the global software industry and as a recipient of rapidly expanding foreign investment.[26] The policy approach adopted by most of these new entrants emulates the highly successful Japanese development model—with significant variations—rather than following traditional Western economic precepts.[27]

---

[24] Alan Wm. Wolff et al., *Conflict among Nations*, p. 528.

[25] See *China and the WTO: Economy at the Crossroads*, Economic Strategy Institute, Washington, D.C., November 1994, pp. 7–8 and *passim*.

[26] Between 1991 and 1993 the amount of direct U.S. investment approved by the Indian government jumped from $104 million to $1.1 billion. John Stremlau, "Dateline Bangalore: Third World Technopolis," *Foreign Policy*, Spring 1996, p. 167. Perhaps more significantly, as a result of the ease of global communications, India has emerged as a major software center. Since 1990, annual software exports soared to $500 million in the 1994–1995 fiscal year. Some estimates expect sales will reach $5 billion annually by 2000. Ibid., p. 153. At the same time, U.S. exports to India were $3.3 billion in 1995, up 43.6 percent from 1994. See *The 1996 National Trade Estimate Report on Foreign Trade Barriers*, Office of the United States Trade Representative (USTR), Washington, D.C., 1996. Nonetheless, competition between American programmers and equally well-trained Indian programmers, paid four times less than their American counterparts, is a new phenomenon with potentially significant economic and political consequences.

[27] Robert Wade, *Governing the Market: Economic Theory and the Role of Government in East Asian Industrialization*, Princeton University Press, Princeton, N.J., 1990.

The commercial policies followed by China illustrate the nature of this challenge. China is the world's fastest growing major economy, with real growth at more than ten percent in 1995 and an average growth rate greater than seven percent for each of the past fourteen years.[28] However, China continues to maintain one of the world's most closed markets for goods and services.[29] A recent report summarizing the challenges the Chinese economic system and its trade policies pose for the multilateral trading system notes that "the Chinese definition of economic reform and [Western concepts] are very different. We mean open markets, they mean limited competition...carefully managed by the government." The report adds that, "[j]udging from recent policy, Chinese leaders view trade as a developmental tool and a way of gaining industrial strength rather than an end in itself. Many have suggested China may follow the model of Japan's trade practices—seeking exports for strategic, industrial gain and shunning mutually beneficial two-way trade. Surely Beijing is aware of Japan's economic success and, like many other governments, would like to emulate it."[30]

Regardless of whether the Chinese are emulating a "Japanese strategy," the Chinese trade surplus with the U.S. is rising rapidly, to $28 billion in 1994 and $34 billion in 1995. Nearly 40 percent of total Chinese exports are to the United States. Interestingly, in addition to textiles and footwear, these exports now include *billions of dollars of electronic machinery, and an ever-increasing volume of higher-value-added products.*[31] China is, in effect, using its market power and growing technical sophistication to create a comparative advantage in targeted high-technology sectors.[32]

---

[28] Statement of Ambassador Michael Kantor, 7 March 1996, before the Senate Foreign Relations Subcommittee on East Asian and Pacific Affairs and the House International Relations Subcommittee on Asia and the Pacific and International Economic Policy and Trade, p. 2.

[29] Ibid.

[30] Economic Strategy Institute, *China and the WTO.* For a description of U.S. trade problems with China (many of which are shared by other industrial countries), see the USTR report *1996 National Trade Estimate Report on Foreign Trade Barriers,* pp. 50–55. U.S. officials cite barriers to imports of computers, medical equipment, heavy machinery, textiles, steel products, chemicals, and pharmaceuticals, and important problems with the enforcement of the 1992 agreement on the protection of intellectual property rights. China's market for services also remains severely restricted. At the same time, China employs a broad range of export subsidies. Statement of Ambassador Michael Kantor, 7 March 1996, p. 6 and *passim.*

[31] Statement of Ambassador Michael Kantor, 7 March 1996, p. 2 (italics added). The United States now runs a deficit with China in electrical machinery.

[32] For a thorough description of China's policy to target microelectronics, including offering market access in exchange for technology transfers, see Thomas Howell, et al., *Semiconductors in China: Defining American Interests,* Semiconductor Industries Association, Washington, D.C., 1995. The report describes the "863 Plan" which aimed to "concentrate...on a few of the most important high-tech fields to catch up with international standards and narrow the gap between China and the world in the next fifteen years." See Zhongguo Keji Luntan No. 5 (September 1990) (JPRS-CST-91-012, cited in Howell et al., p. 55.

The strategy adopted, with its emphasis on exports and its restraints on investments and imports, and the trade problems these policies create are by no means unique to China. Active government programs in Korea, and more recently Taiwan, have been instrumental in the rapid creation of globally competitive advanced-technology industries. Korea is now a leading producer of semiconductors, e.g., Dynamic Random Access Memory chips (DRAMs), with a cost advantage over Japan, and is the second largest consumer, after Japan, of U.S. technology sold as intellectual property.[33] Taiwan has had equally remarkable success. Its national strategy has focused on personal computers and related information products and has made substantial progress toward that goal. Taiwan now holds a leading position as a supplier of a broad range of components and is now the world's largest supplier of CPU boards, monitors, document scanners, graphic cards, mice, keyboards, fax/modems, and most recently laptop computers. Its position is advancing as a supplier of memory chips, logic chip-sets, flat panel displays, and CD ROM drives.[34] These accomplishments are extremely sig-

[33] For a comprehensive review of the Korean miracle, see Alice H. Amsden, *Asia's Next Giant: South Korea and Late Industrialization*, Oxford University Press, New York, 1989. For an excellent discussion of the role and perspective of one of the companies that has achieved a global role in high-technology industry, see the presentation by Y.S. Kim, of the Samsung Electronics Company, to the conference *Sources of International Friction and Cooperation in High-Technology Development and Trade.* Samsung began memory production in 1983 and capitalized on the tremendous growth in demand for DRAMs to emerge as the world's leading supplier of semiconductor memory chips as well as color monitors, with $14 billion in sales in 1994 and 6,000 employees. While government policy certainly helped, this level of success in a fiercely competitive global market involves elements of managerial brilliance, good business practice at every level, and sustained collective effort. For a discussion of the interaction of public policy and private action, see *Competing Economies: America, Europe, and the Pacific Rim*, Office of Technology Assessment, Congress of the United States, Washington, D.C., October 1991, p. 9 and Amsden, *Asia's Next Giant*, pp. 9–10. On U.S. technology purchases, see the National Science Board's *Science and Engineering Indicators*, p. 6-2.

[34] Presentation by Lionel Johns, Office of Science and Technology Policy, The White House, at the National Research Council, 21 May 1996. Government support plays a major role in Taiwan. The successful government program to develop the microelectronics industry was outlined by David C. Hsing, vice president and general director, Industrial Technology Research Institute, Taiwan, at the conference "The U.S., Japan, and the Rules of the Game in High-Technology: National Policies and International Competition in Semiconductors," the Brookings Institution in cooperation with Nomura Institute, 9 May 1996. The program, started in 1974, has proven to be a highly effective model of government-industry cooperation for Taiwan. The program has acquired, developed, and spun off to the private sector a series of fabrication facilities (including the facility, the staff, rights to the technology, and an established market share), each at a more complex level of technology. See also the recent work by Karl J. Fields, *Enterprise and the State in Korea and Taiwan,* Cornell University Press, Ithaca, N.Y., 1995.

nificant. Nonetheless, with respect to China, the scale and growth rate of the Chinese economy, coupled with its current economic strategy and its desire to accede to the World Trade Organization (WTO) as a developing country, pose major challenges to the multilateral trading system.[35]

As this discussion suggests, the level of activity, the number of programs, the substantial funding being made available, the emergence of aggressive new entrants, and most of all the concentration of these programs on similar, if not identical, sectors suggests that the prospect of *increased* international friction is real. **In light of the importance of these national and regional programs and the policy issues they raise, increased and sustained attention should therefore be accorded to these questions by policymakers within national governments and relevant international organizations.**

NATIONAL STRATEGIES:
PRODUCERS VERSUS CONSUMERS

Despite the similarity in long-term goals, there are major differences in national approaches. Some nations are pursuing consumer welfare as an implicit if vaguely defined national goal, while others have adopted quite explicit national economic strategies, designed to pursue national economic strength through producer-oriented policies.[36] A distinctive feature of national policymaking in this latter group is the conscious adoption of a national economic strategy, particularly with respect to high-technology industries. In implementing this strategy, the more producer-oriented countries seek to profit from dynamic opportunities offered by new knowledge-based industries, such as aerospace and semiconductors, and rely on export-led growth to bring their industries down the learning curves that characterize high-technology industries. These producer-oriented countries have enjoyed sustained, high growth

---

[35] Because China is expected to become one of the six biggest traders in the world over the next decade, the policies and practices will have system-wide consequences. See Paul Blustein and Steven Mufson, "A China Trade Question: Is It Ready for Rules?" *The Washington Post,* 19 May 1996, p. H1. The authors, citing Nicholas Lerdy of the Brookings Institution, note that "if you have one major player that doesn't play by market rules and conventions,...then the risk is that the international system will break down...The system can tolerate countries like that if they are very small, but not countries that are such large traders."

[36] Bruce R. Scott, "Economic Strategies of Nations," in Charles Wessner (ed.), *Sources of International Friction and Cooperation in High-Technology Development and Trade.* See also Robert Wade, "Managing Trade: Taiwan and South Korea as Challenges to Economic and Political Science," *Comparative Politics,* vol. 25, no. 2, January 1993, pp. 147–167.

rates.[37] Collectively, these economies, by their scale, rate of growth, increasing regional integration, and impact on the world economy, pose a major challenge to the current international economic order.[38]

The growing impact of these producer-oriented countries on the world trading system makes it important to understand the assumptions, policies, and institutions which some describe as the "developmental state." In broad terms (i.e., with significant variations among nations), the concept of "the developmental state defines a new set of arrangements between the state, society and industry designed to change the structure of the nation's comparative advantage."[39] First developed in Japan, but now imitated, with varying degrees of success in other countries. To a number of analysts, the developmental state has three identifying characteristics: (1) the state plays the role of gatekeeper, determining the terms and conditions for entry of technology, investment, and products into the national economy; (2) there is vertical integration and cross-ownership within large industrial groups; (3) the state has the capability to target key technologies and promote domestic industry—through low-cost capital, the acquisition of foreign technology, and the promotion of lively domestic competition—while encouraging exports. It is important to emphasize that these performance-oriented industrial strategies are quite different from the Western European strategies of

---

[37] *The East Asian Economic Miracle: Economic Growth and Public Policy,* World Bank Policy Research Report, Oxford University Press, New York, 1993. This report outlines the main features of the East Asian economic success story. For the best early analysis of this phenomenon and the role of what he calls the "developmental state," see Chalmers Johnson, *MITI and the Japanese Economic Miracle: The Growth of Industrial Policy, 1925–1975,* Stanford University Press, Stanford, Calif., 1982. See also note 39 below. For an alternative view that places more emphasis on macroeconomic forces in the East Asian record, emphasizing the importance of high savings rates and the rapid diversion of resources from agriculture into manufacturing, see Gary Saxonhouse, "What Is All This About 'Industrial Targeting' in Japan?" *The World Economy,* vol. 6, 1983, pp. 253–273 and Philip Trezise, "Industrial Policy Is Not the Major Reason for Japan's Success," *Brookings Review,* vol. 1, Spring 1983, pp. 13–18.

[38] James Fallows, presentation to the conference *Sources of International Friction and Cooperation in High-Technology Development and Trade, 30–31 May 1995.* See also Fallows, *Looking at the Sun: The Rise of the New East Asian Economic and Political System,* Pantheon Books, New York, 1994, chap. 10, p. 496, note 1 and *passim.*

[39] Citing Chalmers Johnson, Stephen Cohen emphasizes that the rise of the "developmental state" has had a profound effect on international competition. Martin Carnoy, Manuel Castells, Stephen S. Cohen, and Fernando Henrique Cardoso, *The New Global Economy in the Information Age: Reflections on Our Changing World,* Pennsylvania State University, University Park, Pa., 1993, pp. 97–100. For other recent work on related topics, see Thomas M. Huber, *Strategic Economy in Japan,* Westview Press, San Francisco, 1994; Linda Weiss and John M. Hobson, *States and Economic Development: A Comparative Historical Analysis,* Policy Press, Oxford, England, 1995; and Ha-Joon Chang, *The Political Economy of Industrial Policy,* St. Martin's Press, New York, 1994.

the 1960s—sometimes described as "picking losers" because of their focus on ailing national champions.[40]

European national policies were (and are) designed to advance national capabilities in strategic sectors such as computing, communications, and aerospace. Beginning in the 1960s, these national programs focused on the creation of domestic national champions in sectors such as computers. In a number of major European countries, the strategy was based on the premise that larger national firms would have the economies of scale to compete with the large U.S. firms, seen as the principal competitive threat at the time.[41] In pursuit of this concept, European governments created and supported national champions, through a combination of mergers to achieve the necessary scale of operations, direct subsidies, and indirect support provided by favorable national procurement policies.[42]

---

[40] For a discussion of more traditional state aid to industry, see B. Hindley (ed.), *State Investment Companies in Western Europe*, Macmillan Press, London, 1983, *passim*. (See also the section on Dual-Use Technology in Supplement C below.) In the consumer-oriented countries, government decisionmaking is often hampered by a tendency to emphasize political and social factors rather than economic considerations (see OECD Economic Policy Committee, Working Party No. One, *Industrial Subsidies in OECD Countries*: September 1989, p. 49). Unless programs are constructed with appropriate safeguards (e.g., cost sharing), it can also be politically difficult for governments to withdraw from open-ended commitments to support advanced-technology projects. For a discussion of problems with open-ended programs, see Linda Cohen and Roger G. Noll, *The Technology Pork Barrel*, The Brookings Institution, Washington, D.C., 1991. For a discussion of current U.S. partnering programs and safeguards, see Richard J. Brody, *Effective Partnering*.

[41] See, for example, the 1967 study by Jean-Jacques Servan-Schreiber, *Le Defi Americain*, Edition de Noël, Paris. These concerns were shared in Germany: "The German policy for aiding its domestic computer industry grew out of a concern about the technology gap with the United States, and is a means of ensuring an indigenous industry capable of meeting the demands of German companies, hiring German engineers, and exporting to the rest of the world." Michael Kende, "Government Support of the European Information Technology Industry," paper presented at the CEPR-WZB conference held in Berlin 19–20 April 1996, p. 5. The author also cites Brian Murphy, *The International Politics of New Information Technology*, Croom Helm Ltd., United Kingdom, 1986, pp. 65–67.

[42] Michael Kende, "Government Support of the European Information Technology Industry," cites a 1968 U.K. policy on procurement which provided a 25 percent preferential margin over U.S. computer products. Kende adds that both European and U.S. domestic firms benefited from R&D subsidies and government procurement purchases of computer equipment. However, the U.S. policy goal was to create cutting-edge technology for government use, mainly in defense; the U.S. commercial computer industry was a by-product of this policy. In contrast, Kende sees the European policies as largely defensive, with the goal of creating a computer industry per se. This reflects a rather static view of IBM's success—that size leads to success, rather than the more dynamic view that success leads to size. This led to a focus on creating large firms rather than firms with competitive products. Kende attributes the failure

While the record is mixed, these national policies often altered market outcomes, albeit at considerable cost to the public purse. For example, in computers, the national champion policies largely failed, though the subsidies and favorable procurement practices ensured a market share for national champions, such as Machines Bull, to the detriment of foreign and any remaining domestic competitors. In some cases, these nationally based policies were replaced by regional efforts in the 1970s. The example of Airbus, a major regional success, is perhaps paramount. Though currently much emphasis is placed on European regional approaches, in many cases national efforts to develop or sustain national champions have continued. Indeed, some of the national efforts of the 1980s, though expensive, had considerable technical success, especially in large projects or where procurement could play a supportive role.[43]

For example, in the view of some analysts, French national development programs have generally proven most effective when the objective required large-scale mobilization of resources, when the number of technological results could be limited, and when competitive market forces could be suppressed or contained by the state.[44] French technological successes under this strategy include participation in the Ariane launch program and Airbus, as well as the TGV (the high speed train), the national nuclear energy program, and the rapid modernization of the telephone system. The centralized French approach has proved much less successful in circumstances where a company is required to rapidly adapt its products and processes to changing international market conditions.[45]

---

of this policy to the lack of attention to *technology* (rather than firm size) and to competitive applications of *existing technologies,* as well as to the absence of a supportive policy framework for entrepreneurs. *Passim.* (Government procurement continues to be a source of aid for national firms. See Supplement C.) Kende also argues that European governments continue to give insufficient attention to the creation of a policy framework conducive to the development of innovative, entrepreneurial firms. Ibid., pp. 26–29.

[43] See Douglas Webber, "Alcatel, Francetelecom, and the French Government," in Olivier Cadot et al., *European Casebook on Industrial and Trade Policy,* Prentice Hall, New York, 1996, pp. 219–237. In a similar approach, current U.S. government policy is to encourage rational downsizing of the defense industry through the reimbursement of some restructuring costs incurred as a result of mergers and acquisitions. See Lawrence J. Korb, "Military Metamorphosis," *Issues in Science and Technology,* Winter 1995–1996, pp. 75–77.

[44] For an elaboration of this analysis, see John Zysman, *Political Strategies for Industrial Order: State, Market and Industry in France,* University of California Press, Berkeley, 1977.

[45] Ibid.

THE IMPORTANCE OF CONDITIONAL
GOVERNMENT SUPPORT

A key element in the success of these government programs may be the degree to which the government can make its support conditional.[46] Recognizing the deficiencies of previous "no strings" aid to national champions, the early Mitterand government in France negotiated extensive performance requirements with corporate recipients of state aid.[47] Similarly, to some observers of the East Asian experience, a distinguishing feature of East Asian practice is the provision of *conditional government support,* with the expectation that investment and export targets will be met by domestic firms benefiting from a broad array of government support.[48]

The industrialization of Korea, and particularly its success in moving into high-technology, high-value-added industries, has been especially remarkable. Some analysts attribute this to the terms and conditions the government has been able to impose on industry in exchange for preferential national and international credits. "Throughout most of the 25 years of Korean industrial expansion, long-term credit has been allocated by the government to selected firms at negative real interest rates in order to stimulate specific industries."[49] However, the government also imposed performance standards, which included export and investment targets as well as price controls, restrictions on capacity expansions, limits on market entry, prohibitions on capital flight, and restraints on tax evasion buttressed by government control over the banking system. In this view, the foundation of late industrialization is the subsidy (defined to include both protection and financial incentives), but a crucial condition for the success of these policies

---

[46] It is important to underscore that other factors such as tax policy and legal and financial institutions play a critical role in the performance of individual firms. For example, the policy framework in some regions of the United States supports an entrepreneurial culture through the interaction of a well-developed system of higher education, a highly mobile skilled labor force, and networks of local specialized suppliers and venture capitalists, plus the well-developed national capital markets such as the NASDAQ. This policy framework is further strengthened by a business ethic which places relatively low social penalties on failure.

[47] See D. Webber in Cadot et al., *European Casebook.* Alcatel is a case where an aggressive, company-led policy of mergers and acquisitions, bolstered by grants, state-supported R&D, and massive public procurement resulted in both world-class technologies and a strong, internationally competitive position in telecommunications equipment.

[48] Alice H. Amsden, *Asia's Next Giant,* chap. 6, especially pp. 143–146. See also Fallows, *Looking into the Sun,* p. 445 and p. 497, note 2. Fallows also cites Amsden, *Diffusion of Development: The Late Industrializing Model and Greater East Asia,* Papers and Proceedings of the American Economic Association, May 1991, pp. 284–285. See also the World Bank Policy Research Report, *The East Asian Economic Miracle: Economic Growth and Public Policy.*

[49] Amsden, *Asia's Next Giant,* p. 144.

is a government sufficiently strong politically "to impose performance standards on the interest groups receiving public support."[50]

THE IMPORTANCE OF SUSTAINED EFFORT

These policies have yielded results. Indeed, to a considerable degree, the emerging Asian countries have proven more successful, in specific sectors, in developing technologically advanced industries able to compete in rapidly evolving global markets than have some countries with more established technological infrastructures. Importantly, much of the East Asian success has involved the emulation and effective commercialization of *existing* technologies, rather than efforts to develop *new* technologies, while the latter approach has characterized a number of European regional programs.[51]

The record of government policies to develop new technologies and support high-technology industry is by no means one of endless success, in East Asia or elsewhere. In the United States, national security and government mission programs have had remarkable success, sometimes across entire sectors, as well as notable failures. The national champion strategies of France and Germany have encountered failures and successes, and now include highly successful regional efforts. The sector-specific, multifirm strategies adopted in East Asia, with pervasive support by governments eager to capture the benefits of advanced industries, have had considerable success, though this achievement is also the result of great collective effort

---

[50] Ibid., p. 145. In the case of Korea, government discipline in terms of its vigorous insistence on meeting export targets, regardless of political connections, had major positive implications for efficiency. These pressures were backed by (1) the government's control of the banking system; (2) limitations on the number of firms per industry; (3) negotiated price controls to curb monopoly power; (4) controls on capital flight; and (5) taxation of the middle classes with limited social services available to the lower classes. Amsden, *Asia's Next Giant,* pp. 16–18.

[51] While it is beyond the scope of this study to compare these programs, some observers have noted that no European firm has matched the success of American or Asian firms in assembling and marketing IBM-compatible personal computers based on widely available technologies. Kende, "Government Support of the European Information Technology Industry," pp. 20–21. European policymakers increasingly recognize the importance of these factors. A "Green Paper on innovation" recently issued by the European Commission notes that "one of the great paradoxes of the European Union is that, despite its internationally acknowledged scientific excellence, it launches fewer new products, services and processes than its main competitors," adding that "this state of affairs results from structural obstacles such as a complex legal and administrative environment, unsuitable financing systems, etc." See *RTD Info,* European Commission for Science, Research, and Development, Brussels, Belgium, February 1996.

in a favorable international environment.[52] **The common thread in each of these cases is that policymakers accept possible setbacks, even failure, as a fact of life when investing in new technologies. To a considerable extent they share a common goal. Most importantly, they recognize that the international market standard is not perfection, but rather the ability over time to make better decisions than their competitors.**[53]

Making better decisions than one's competitors is not self-evident, for either private or public investors. Moreover, investments in R&D are inherently risky. Consequently, some government-supported R&D efforts, like those in the private sector, will be unsuccessful. (Indeed, the absence of *any* failure would suggest excess caution.) But as noted above, successful R&D investments can change, and have changed, the capability—and the competitive position—of an entire economy. For example, in the United States "government support was crucial in areas such as computers and integrated circuits, jet engines and airframes, and biotechnology and medical equipment. The result has been entire fields of productive wealth-enhancing, job-creating economic activity."[54] Japan has achieved similar success, albeit with a different approach.[55]

---

[52] The current international environment is conducive to rapid industrialization in the sense that late-industrializing countries have a backlog of technologies to draw on and, in industries such as electronics, benefit from a rapidly growing market which eases the way for new entrants. Relatively open markets, in the United States and Europe for example, and access to advanced manufacturing equipment also make rapid industrial growth possible in high-technology industries. The process remains difficult, for both private actors and public policymakers. Amsden highlights the difficult trade-offs faced by policymakers in late-industrializing countries, e.g., the need for low interest rates to stimulate investment, and high rates to encourage savings; for undervalued exchange rates to boost exports, and overvalued exchange rates to lower the cost of foreign debt and imports; and for protection from foreign competition, but a free trade environment for exports and access to imports of capital equipment. Amsden, *Asia's Next Giant,* p. 13.

[53] Bruce Scott, personal communication with National Research Council, March 1995. A recent OECD study makes a similar point, noting that "governments' track record in picking winners is not good, but that a very few winners may be worth many losers." Martin Brown, *Impacts of National Technology Programs,* OECD, Paris, 1995, p. 21.

[54] Council of Economic Advisers, *Supporting Research and Development to Promote Economic Growth,* p. 8.

[55] For example, Japanese efforts to develop their national semiconductor industry in the 1970s achieved great success. See William J. Spencer, *SEMATECH, 20* November 1995 contribution to Steering Committee deliberations. The author cites the highly successful Japanese VLSI project focused on dynamic random access memory (DRAM) development and manufacture, initiated in 1975 at a cost of about $300 million. In four years, the program helped bring the Japanese industry "from a small player in the total semiconductor business to become the dominant semiconductor producer by the mid-1980s." P. 3. For an excellent recent discussion of the rise of the Japanese semiconductor industry, see Kenneth Flamm, *Misman-*

EXPORT-ORIENTED ECONOMIES

Another relevant feature of producer-oriented countries is their active promotion of exports. Their policymakers are interested in specific market outcomes, not just maintaining the rules of the trade regime and ensuring their proper enforcement. Typically, these countries vest more power, (i.e., the ability to claim scarce resources) in the hands of producer institutions—at the expense of their consumers—and thus mobilize a higher fraction of their incomes for productive purposes than is the norm among the established industrialized countries. And, as noted above, the government emphasizes exports as a means both of encouraging efficiency in targeted industries and of repaying the external debt used to finance their expansion. In consequence, this policy orientation significantly enhances the export capacity of these economies.[56]

Moreover, the impact of these different assumptions and objectives is compounded by the tendency of the consumer-oriented economies, such as the U.S. and western Europe, to allocate resources to address income inequality,[57] whereas producer economies, especially those aided by a supportive social system and more even income distribution, are allocating resources and following export-oriented policies to capture the knowledge-

---

aged Trade? especially chap. 2. Flamm emphasizes the importance of the VLSI project, describing it as the "largest infusion of R&D subsidies ever received by the Japanese semiconductor industry in both absolute and relative terms." P. 96. (See also Box F below.) For a discussion of the national objectives of Japan—and other nations—in the aerospace industry, see Box H.

[56] Bruce Scott, *Economic Strategies of Nations*, p. 27. Some analysts see the impact of national policies, especially coherent national strategies, as an increasingly important element in global competition. In this view, the implicit competition between the North Atlantic area and East Asia is not a locational competition based on shifting comparative advantage. "To a degree, it is a competition between differing economic strategies . . . with the producer economies" focused on achieving higher growth as a way to enhance their economic and political power, not their short-term standard of living. Global economic competition is, therefore, in part between neo-classical strategies focused on short-term consumer welfare and neo-mercantilist strategies focused on development of economic power." Ibid. See Robert Wade, "Managing Trade," Robert Wade, *Governing the Market,* and Amsden, *Asia's Next Giant.* See also James Fallows, *Looking at the Sun,* Chalmers Johnson, *MITI and the Japanese Economic Miracle,* Stephen Cohen and Pei-Hsiung Chin, *Tipping the Balance,* and Thomas Huber, *Strategic Economy in Japan.* See also Robert A. Blecker, *Beyond the Twin Deficits,* M.E. Sharpe, New York, 1992, chap. 5 and 6.

[57] This tendency seems to be changing. In many countries, social spending is falling in real terms (as a percentage of GNP share) in the 1990s.

intensive industries that they believe offer the best prospects for future growth and income-generating employment.[58]

In addition, these industries are seen as strategic because they provide inputs that greatly enhance productivity throughout other sectors of the economy. For example, the electronics and materials industries provide critical inputs for downstream industries. These inputs are essential for the international competitiveness of the user industries. Moreover, the development of the strategic industries themselves is critically dependent on their relationships with the downstream industries which employ their products as production inputs.

Because high-technology industries are believed to have unique characteristics (which can make intervention effective) and tremendous growth potential (which makes investments worthwhile), high-technology industries are the target of the industrial policies of many participants in the multilateral trading system. The national practitioners of these promotional policies, i.e., those who hold responsibility for national economic development, do not accept the logic of free trade theorists, or even moderate trade policy practitioners. Instead, they are convinced that high-technology industries are strategic in terms of their impact on economic growth and national autonomy.[59] (See Box A.) In consequence, they bring to bear a host of policy measures to protect and promote these industries. (See Box B.) A number of these countries also have divergent policy preferences and national traditions with respect to intellectual property protection, open investment regimes, and competition policy. (Each of these issues is discussed below.) In combination, these anticompetitive practices, policies, and traditions can give substantial economic advantages to the nations that directly support strategic industries.[60]

---

[58] Bruce R. Scott, presentation to the conference *Sources of International Friction and Cooperation in High-Technology Development and Trade, 30–31 May 1995.* See also Scott, *Economic Strategies of Nations, op. cit.,* pp. 42–49. Scott notes that under these policies, consumers are forced to subsidize producers to accelerate capital formation, but if the national strategy is successful, consumers also gain enormously, in the medium term, as their incomes rise dramatically. Amsden, *Asia's Next Giant,* emphasizes that although the Korean government has spent more than it has collected, "*it has spent more on long-term investment, not on short-term consumption.*" (Italics in original), p. 92.

[59] Stephen Cohen and Pei-Hsiung Chin, *Tipping the Balance: Trade Conflicts and the Necessity of Managed Competition in Strategic Industries,* paper delivered at Kiel Conference, *Towards a New Global Framework for High-Technology Competition, 30–31 August 1995.* The authors argue that these industries are strategic in every sense that a government might understand: that is, in terms of their ultimate impact on power, wealth, and culture. The same dynamic characteristics of these industries that defy normal market equilibria also give leverage and consequence to government intervention.

[60] Laura Tyson, *Who's Bashing Whom: Trade Conflict in High-Technology Industries,* Institute for International Economics, Washington, D.C., 1992, p. 45. See also the recent OECD review by Martin Brown, *Impacts of National Technology Programs,* Paris, 1995, p. 21.

Together these policies pose a philosophical and practical challenge to the current trading regime. Government support for selected industries, and especially high-technology industries with their perceived potential for rapid growth, is philosophically at odds with the principles of open international competition. As a recent study of U.S.-Japanese competition in semiconductors observes, "the post-war trading system aspired to an ideal of free competition among firms from all nations operating in a single, open, global market, with market outcomes determined only by the efficiency and effectiveness of individual companies. How can this vision be reconciled with the realities of a world in which national governments make large investments in new technologies, which may totally alter the industrial landscape, and inevitably favor those firms with the easiest access to the innovations created with these subsidies."[61] While some would argue that in the past these principles were often honored in the breach, the practical implications of the current level of government intervention, the increasing number of governments practicing such intervention, and the concentration of government support on a limited range of sectors make the challenge to the current multilateral trading system and its supporting assumptions an increasingly critical issue for policymakers.

BOX A. WHY ARE COUNTRIES CONCERNED ABOUT THEIR HIGH-TECHNOLOGY INDUSTRIES?

Throughout this report, attention is focused on firms that develop and produce advanced technological products. As noted above, not all economists accept the view that high-technology industries are significantly different from traditional industries (potato chips versus computer chips) and therefore deserving of greater attention from policymakers. There is, however, a growing body of economic thought that argues that the composition of the economy matters and that high-technology industries bring special benefits to national economies.[62]

The benefits attributed to high-technology industries rest on a number of interlocking observations.

continued

---

[61] Kenneth Flamm, *Mismanaged Trade*, p. 3.

[62] For a summary of the new trade and growth theory, see Luc Soete, "Technology Policy and the International Trading System: Where Do we Stand?"; paper presented at the conference Towards a New Global Framework for High-Technology Competition, 30–31 August 1995, Kiel, Germany. For an early exposition of the strategic trade argument, see James Brander and Barbara Spencer, "Export Subsidies and International Market Share Rivalry," Journal of International Economics, February 1985. For an overview of strategic trade theory and its increasing attraction to national governments, see Jeffrey A. Hart and Aseem Prakash, "Implications of Strategic Trade for the World Economic Order," paper prepared for the Annual Meeting of the International Studies Association, San Diego, Calif., 16–20 April 1996. For a critique of the strategic trade concept, see Paul Krugman, Peddling Prosperity, W.W. Norton Press, New York, 1994, pp. 239–244 and chap.10.

First, **high-technology firms are associated with innovation**. Firms that are innova-tive tend to gain market share, create new product markets, and use resources more productively. This proposition is supported by the findings of a recent National Research Council conference on the impact of innovation on productivity, wages, and employment.[63]

Second, **high-technology firms perform larger amounts of R&D than more traditional industries**. High-technology firms are identified by the very high percent-age of their revenue devoted to research—often more than 10 percent—as compared with a 3 percent level for more traditional industries. Collectively, high-technology industries constitute a disproportionate share of total private R&D spending in the U.S. And the social returns of such R&D spending are widely believed to far exceed the private returns.[64]

Third, **these positive spillover effects benefit other commercial sectors by generating new products and processes that can lead to productivity gains and generate new manufacturing opportunities**.[65] Advances in electronics have made it a key enabling industry responsible for new methods of manufacturing in steel, automobiles, aerospace, and even agriculture, as well as the creation of a whole gamut of consumer electronic and defense related products. There is substantial economic literature underscoring the high returns of technological innovation, with private inno-vators obtaining a rate of return in the 20 to 30 percent range with the spillover (or social return) averaging about 50 percent.[66]

Fourth, **the positive spillover effects are often locally concentrated**. Firms frequently concentrate in particular locations to benefit from the externalities associ-ated with a qualified labor supply with appropriate skills, specialized suppliers of inputs and supporting services, and informal horizontal information networks for the exchange of the "tacit" knowledge required for the exploitation of new techniques and processes. These "network systems flourish in regional agglomerations where repeated interaction builds shared identities and mutual trust while at the same time intensifying rivalries."[67]

---

[63] National Research Council conference *Technology, Wages, Productivity, and Employ-ment, 1–2 May 1995, Conference Proceedings* (forthcoming). See also Gregory Tassey, *Tech-nology and Economic Growth: Implications for Federal Policy,* NIST Planning Report 95-3, U.S. Department of Commerce, Washington, D.C., 1995, p. 12.

[64] Martin N. Baily, and A. Chakrabarti, *Innovation and the Productivity Crisis.* The Brookings Institution, Washington, D.C., 1988, and Zvi Griliches, *The Search for R&D Spillovers,* Harvard University, Cambridge, Mass., 1990.

[65] Lawrence M. Rausch, *Asia's New High-Tech Competitors,* NSF 95-309, National Sci-ence Foundation, Arlington, Va. 1995.

[66] Ishaq Nadiri, *Innovations and Technological Spillovers,* NBER Working Paper No. 4423, 1993, and Edwin Mansfield, "Academic Research and Industrial Innovation," *Research Policy,* February 1991. See also Council of Economic Advisers, *Supporting Research and Develop-ment to Promote Economic Growth.*

[67] Annalee Saxenian, *Regional Advantage: Culture and Competition in Silicon Valley and Route 128,* Harvard University Press, Cambridge, Mass, 1994, p. 4. The author identifies two broad types of industrial systems: independent firm-based systems typically associated with capital-intensive industries, such as oil and automobiles, versus network-based systems, orga-nized around horizontal networks of firms where producers deepen their own specialized capa-bilities while cooperating with other specialists. See also Nitin Nohria and Robert G. Eccles (eds.), *Networks and Organizations: Structure, Form, and Action,* Harvard Business School Press, Boston, Mass., 1992.

Because these local externalities tend to be self-reinforcing, the competitive position of the relevant industry tends to improve over time. Conversely, the decline in an industry's position tends to erode the specialized infrastructure as well.[68]

Fifth, **high-technology products are a major source of national economic growth in all of the major industrialized countries,** because the global market for high-technology manufactured goods is growing at a faster rate than are the markets for other manufactured goods.[69] For example, in the United States, sectors such as aerospace, information systems (software, computers, and semiconductors), chemicals, pharmaceuticals, biotechnology, and medical equipment are all leading sources of U.S. exports. Moreover, as noted above, these high-technology industries also account for a disproportionate amount of total industrial R&D.

Sixth, as one would expect from the above, **high-technology firms are associated with high value-added manufacturing and, importantly, the creation of high wage employment.** The firms that innovate rapidly, introduce new technologies, develop new products, and expand exports are also the firms that increase employment and contribute disproportionately to the national R&D effort.[70]

Seventh, **many high-technology industries have important consequences for core government missions.** Foremost among these is national defense. Early, assured access to advanced, low-cost technologies is viewed by many as a critical element in a viable defense strategy for the next century.[71] As one informed observer remarked, without technological superiority, military superiority becomes a question of numbers and training.[72] The impact of new enabling technologies can be equally crucial for major government missions in energy development, environmental protection, and health care (where new technologies offer major advances in methods, drugs, devices, and equipment).

---

[68] Jay Stowsky, "Regional Histories and the Cycle of Industrial Innovation: A Review of Some Recent Literature," *Berkeley Planning Journal,* vol. 4, 1989; Michael Porter, *The Competitive Advantage of Nations,* Free Press, New York, 1990; and Lester Thurow, *Head to Head: The Coming Economic Battle Among Japan, Europe, and America,* Morrow, New York, 1992.

[69] See 1993 Science and Engineering Indicators, National Science Foundation, Arlington, Va., 1993, pp. 159–160 and chap. 6. For example, in constant 1980 dollars, one recent study found that "production of high-tech manufactures by the major industrialized nations more than doubled from 1981 to 1992, while production of other manufactured goods grew by just 29%." In 1981, U.S. high-technology manufactures represented 15 percent of total U.S. manufactured output. By 1992, this figure rose to an estimated 27 percent. In Japan in 1981, high-technology manufactures represented nearly 17 percent of total Japanese production; the figure in 1992 was over 31 percent. For European Community members, the comparable figures were 12 percent in 1981 and 17 percent in 1992. See also the *1993 U.S. Industrial Outlook,* International Trade Administration, Washington, D.C., 1993, p. 21; and Lawrence Rausch, *Asia's New High-Tech Competitors,* p. 7.

[70] Laura Tyson, *Who's Bashing Whom?* p. 32. It is also true that, in some cases, the introduction of new technologies may displace some workers by making certain types of skills obsolete. This can lead to structural unemployment through a mismatch between the skills of the workers displaced and the (new) skills demanded by the marketplace.

[71] See Flat Panel Display Task Force, *Building U.S. Capabilities in Flat Panel Displays: Final Report,* U.S. Department of Defense, Washington, D.C., October 1994.

[72] Rear Admiral Marc Y.E. Pelaez, Chief of Naval Research, is the source of this observation. See C. Wessner (ed.), *Sources of International Friction and Cooperation in High-Technology Development and Trade.*

CREATING COMPARATIVE ADVANTAGE

Collectively, these policies can be powerful tools for *creating* comparative advantage. This is because intervention can change the operation of the market in these highly dynamic, learning-based industries, resulting in long-term sectoral advantage. (See Box F.) Given that many of these sectors represent large markets, often experiencing significant growth, the economic benefits to be garnered by successful government interventions can be substantial. Moreover, success in one strategic industry can also enhance a nation's capacity to acquire other technologies and achieve other national objectives in the international arena.[73]

In short, many governments in the industrial and industrializing world believe there is tremendous growth inherent in many high-technology industries and often see these industries as strategic national assets as well. "The payoff is to national governments from such investments is that the local economy can become internationally competitive while maintaining national autonomy and—more important—such investments lead to higher levels of local productivity, providing greater political space for any government."[74] Private companies share the government's interest in supporting these sectors because these policies generate demand for their more sophisticated products while increasing the supply of more highly skilled technicians, workers, and scientists—which in turn allows national corporations to be more competitive globally.[75]

Whatever the rationale, it is clear that there are genuine differences in national attitudes concerning a nation's knowledge and technology base. Industrial policy in Japan, for example, is designed to maintain and enhance the nation's technological capability, a matter which is normally not a subject of ongoing, integrated policy concern within the United States, except for defense needs or when the results of the absence of policy become apparent.[76] Though belated, U.S. policy responses, in key areas such as

---

[73] Stephen Cohen, *Tipping the Balance,* p. 5. For a broader, penetrating analysis of the relationships between American economic strength, national goals, and the current international political system, see Michael Borrus, Wayne Sandholtz, John Zysman, Ken Conca, Jay Stowsky, Steven Vogel, and Steve Weber, *The Highest Stakes: The Economic Foundations of the Next Security System,* Oxford University Press, New York, 1992, *passim.*

[74] Martin Carnoy, "Multinationals in a Changing World Economy: Whither the Nation-State?"; chap. 3 in Carnoy et al., *The New Global Economy,* p. 89.

[75] Ibid.

[76] See the presentation by John P. Stern, "Japan: The Philosophy of Government Support for Information Technology," in Charles Wessner (ed.*), Symposium on International Access to National Technology Programs, 19 January 1996.* National Academy Press, Washington, D.C., forthcoming.

semiconductors, have proved successful.[77] Reflecting a relative decline in key technologies such as electronics, Europeans have launched—with mixed success—a broad range of programs, at both the national and Community levels, intended to redress their competitive position.[78] Notwithstanding these similarities, there are genuine differences in perspective.

The differences in perspectives are perhaps most evident between the United States and Japan. Two American analysts characterized these differences as follows:

> Japan, we believe, values industries differently than does America...[and believes] that industries have importance beyond the goods they produce. Acting on this belief, the Japanese are driven to procure or develop skills and knowledge that they may lack for their domestic economy so that non-production benefits—especially learning and diffusion—can be realized at home. Industrial policy in Japan is guided by the effort to maintain the nation's knowledge and technology base rather than to produce a specific product to which a domestic firm might affix a nameplate... The U.S., in contrast, does not value industries in this way,... leading to wholesale capacity losses, or even domestic skill displacement from the American economy that Japan would never tolerate... As we have seen in the aircraft industry, Japan is willing to pay (and pay dearly) for the same technical knowledge that the U.S. is willing to transfer abroad because it values the ancillary industrial results of that knowledge as much [as], or more than, the ability to make specific goods.[79]

Whether it is the perception that these industries are strategic, or merely crucial opportunities for employment and growth, the combination of these factors gives powerful support to the contention that strategic and locational competition, based on a wide array of diverse policy measures, is likely to accelerate for the foreseeable future.

---

[77] The increase in the global market share of U.S. semiconductor firms as compared with Japanese firms is frequently cited as a sign of the resurgence of the U.S. industry. Stern notes that the 1993 U.S. share of the semiconductor market was 2 percent larger than the Japanese share. While this is an improvement, from the U.S. perspective, over the 10 percent lead the Japanese industry enjoyed in 1990, it is "nowhere near the 53 percent of the world market" held by the U.S. semiconductor industry in 1984. Ibid, p. 3.

[78] See the European Commission, *European Report on Science and Technology Indicators, 1994.* For a discussion of different approaches to industrial policy in Europe, Japan, and the United States, see Jeffrey A. Hart, *Rival Capitalists: International Competitiveness in the United States, Japan, and Western Europe*, Cornell University Press, Ithaca, N.Y., 1992.

[79] David Friedman and Richard Samuels cited in Charles H. Fine and Daniel E. Whitney, "Is the Make-Buy Decision Process a Core Competence?"; MIT Center for Technology, Policy, and Industrial Development, Cambridge, Mass., January 1996, p. 6. For a broader discussion of the longstanding Japanese approach to the acquisition or indigenization, diffusion, and nurturing of technology as a means of ensuring national technological capabilities that enhance that nation's security, see Richard Samuels, *Rich Nation, Strong Army: National Security and the Technological Transformation of Japan,* Cornell University Press, Ithaca, N.Y., 1994.

INTERNATIONALIZATION OF "DOMESTIC" POLICIES

In this context, seemingly domestic concerns about how best to support innovation, sustain the development of science and technology for industrial productivity, and promote advanced technologies become issues on the forefront of the international political and economic agenda.[80] Indeed, individually, the policy components for technological development are already contentious international issues. "There are disputes over the reach of intellectual property rights, over dumping practices, and subsidies. When taken together, they raise an international debate about national development, about how to generate and retain advantage in the technologies and industries on which future development and security will rest."[81] In this view, the policies, instruments, and practices which have traditionally been considered domestic prerogatives—especially in conjunction with "traditional" trade barriers—could become the basis for more serious conflict among the principal regions of the global economy.

Trade barriers are already a major source of friction in the international system. The "normal" frictions resulting from different national practices and uneven implementation of international accords already require sustained, professional attention. It may be that the task of resolving, or containing, trade disputes will become steadily more difficult as a result of the growth in trade, foreign direct investment, and the accompanying interpenetration of national economies. Trade disputes in high-technology sectors have proved persistent, being difficult to resolve with any finality, especially when the strategic economic interests of trading partners are engaged.

These high-technology trade frictions emerge, at least in part, because governments employ a wide variety of policy measures designed to support high-technology industries. Indeed, one of the explanations for the recurrence of market access frictions in technology-intensive industries lies in the fact that, for many governments, trade and investment in these industries take on a strategic meaning.[82] Consequently, the competitive process in such industries is affected by a host of formal and informal national policies. Despite a clear worldwide trend toward liberalization of trade and investment regimes over the last decade, national governments have throughout this period revealed a clear preference for maintaining the ability to support, attract, or retain high-technology producers in their territories.[83] International friction may well increase in frequency and intensity as a growing number of nations compete for what they perceive to be the technologies and industries of the future.

---

[80] See Michael Borrus et al., *The Highest Stakes,* p. 178. See also Sylvia Ostry and Richard Nelson, *Techno-Nationalism and Techno-Globalism,* preface, pp. xvii–xviii, and USTR, *1996 National Trade Estimate Report on Foreign Trade Barriers.*

[81] Borrus et al., *The Highest Stakes,* p. 178.

[82] *Market Access and Competition in Technology-Intensive Industries: Issues Paper.* OECD, Paris, 26–27 October 1995, p. 4.

[83] Ibid.

BOX B. HOW DO GOVERNMENTS SUPPORT HIGH-TECHNOLOGY INDUSTRIES?

Governments support high-technology industries through a vast array of policy measures, often addressing seemingly quite disparate policy objectives. Trade-related measures continue to play a central role. Though trade measures elude fixed definitions, they include a panoply of laws, regulations, policies, and practices that protect domestic products and markets from foreign competition or stimulate exports of selected domestic products.[84] A recent U.S. government report identified nine different categories of government measures that "restrict, prevent, or impede" international commerce.

These categories include **restrictive import policies,** such as tariffs, quotas, import licensing, and customs barriers; **standards, testing, labeling, and certification; government procurement,** such as "buy national" policies or practices (see Supplement E); **export subsidies; the lack of intellectual property protection** as a result of inadequate patent, copyright, and trademark regimes; **services barriers; investment barriers** involving limitations on foreign equity participation and on access to government-funded R&D programs and other restrictions; and **anticompetitive practices** with trade effects which are tolerated or encouraged by governments.[85]

Other common policies, which are either designed, or provide the opportunity, to improve the competitiveness of national firms, include deregulation, privatization, relaxation of product and environmental standards, encouragement of mergers and strategic alliances, and targeted tax measures designed to encourage innovation and investment.

Governments also support high-technology industry under an exceedingly broad range of policy objectives and implementing financial instruments. For almost a decade, the Organization for Economic Cooperation and Development (OECD) has sought to assemble detailed internationally comparable data on national support to industry.[86]

On the basis of this analysis, expenditure by OECD member countries was in excess of $66 billion per annum in the period 1986–1989. The total amount spent on these measures declined in that period, primarily because of an overall reduction in targeted tax expenditure. However, government support is increasingly focused, with greater use of direct grants, government guarantees, and support for exports for selected industries. While regular and comparable reporting on subsidies to manufacturing industries does not yet exist, the OECD work captures the scope and diversity of national policy objectives and instruments.

Policy Objectives

In addition to support for investment in particular high-technology sectors, such as microelectronics, biotechnology, and aerospace, governments often pursue what the OECD describes as "horizontal objectives" such as support for regional development,

continued

---

[84] USTR, *1996 National Trade Estimate Report on Foreign Trade Barriers,* p. 1.

[85] Ibid, pp. 1–2.

[86] The OECD exercise has succeeded in creating three major assets: a database of national support programs for industry with calculations of "net subsidies" for each measure, a "peer review" mechanism for collectively reviewing the data submitted, and an improved understanding of the scope and diversity of government support mechanisms. For a discussion of different types of national technology support programs, see Martin Brown, *Impacts of National Technology Programmes,* chap. 2.

aid to small and medium-size enterprises, aid for employment and training, support for "enterprises in difficulty," export incentives, and other trade-related assistance, as well as support for research and development. R&D support includes both traditional government support for research through grants to universities and research institutes and support to industry, either through direct subsidies (e.g., in the electrical and aerospace sectors) or more indirectly through defense contracts. Data collection has recently expanded to include government support for energy efficiency and for environmental protection. Currently, more than 1500 programs and measures of support are available to manufacturing in OECD countries.[87]

Instruments

Financial instruments include direct grants to companies; preferential loans; government guarantees for loans; equity capital infusions by government entities or government-controlled banks (often to cover recurrent losses); preferential government procurement policies, and targeted tax concessions for specific sectors, for "underdeveloped" regions (where high-technology industries may be located), and for specific activities, such as R&D. Government goods and services (e.g., electricity) are also provided at below cost, and domestic industries (especially when they are state-owned or -controlled) can be required to purchase domestic products (e.g., electric turbines or telecommunications equipment) at prices exceeding those available on world markets, thereby providing an important source of funding for other activities such as R&D investment or export support.

Government policy guidance and the activities of its agencies also contribute to the support of high-technology industry through policies such as government sponsorship (with or without financial contributions) of research consortia, selective antitrust exemptions for joint R&D efforts and cooperative production arrangements (important in countries where there is not a systematic failure to enforce antitrust policies), transfer to industry of intellectual property resulting from government-financed research in countries with adequate and effective intellectual property protection, the transfer of defense-related technology for civilian use, the setting of industrial standards, and the design of rules of origin.

Governments use the above policy instruments in different combinations. Indeed, these measures are most effective in combination, as part of an integrated strategy to support a particular industry. Countries deploy these measures differently, reflecting historical differences in systems of corporate governance, levels of direct state intervention in the economy, and relative openness of the national economy toward foreign investment and imports. Whatever the rationale and policy guise under which these measures are deployed, "they all aim at the objective of spurring the development of new technologies by domestic industries and strengthening the competitive position of national (or domestically established) firms."[88]

---

[87] OECD, *Industrial Subsidies: A Reporting Manual,* Paris, 1995.

[88] OECD, *Market Access and Competition in Technology-Intensive Industries.* In addition to discussing targeted measures of the type described above, the OECD Issues Paper underscores, inter alia, the importance of market access, rules of origin, quantitative restrictions of imports, performance requirements, and patent protection, noting that this set of trade practices is a significant source of trade friction, not least because these measures are actively used to

NATIONAL LOCATIONAL COMPETITION
AND ITS IMPACT ON SCIENTIFIC
RESEARCH AND COOPERATION

Strategic competition will remain a source of tension among nations and may threaten international norms in areas such as international scientific cooperation for the development of new technologies and for the maintenance of the multilateral trading system, particularly with respect to trade in high-technology products.[89]

The risk of negative impact on public support (i.e., funding) for basic research and on the willingness to participate in international scientific cooperation is believed to be real, although this area requires further analysis.[90] Pressures for closer government direction and control over publicly-supported science have grown from two sources: (1) the pressure of government deficits; and (2) the specific concerns of politicians and administrators that the scientific establishment has not been addressing national needs. This has resulted in pressure to work on applied projects that would bring the work of scientific researchers into closer connection with market-oriented, industrial R&D projects.

Some anecdotal evidence suggests that a decline in scientific cooperation in some commercially relevant disciplines may already be present. For example, it is increasingly common for scientific publications to not fully disclose relevant findings that would permit replication by other scientists and for restrictions to be adopted on access to "intermediary results" for research-related material that cannot be published in journals, such as experimental materials, innovative instruments, software, and data sets.[91] Moreover, the traditional emphasis on cooperation in basic research may no longer have the same applicability, particularly in those sectors such as computer science or biotechnology where the basic/applied dichotomy has eroded.[92]

---

build national or regional production bases in high-technology industries. These issues are taken up in the discussion below.

[89] Stephen Cohen and Pei-Hsiung Chin, *Tipping the Balance, passim.*

[90] Partha Dasgupta and Paul A. David, "Toward a New Economics of Science," *Research Policy,* vol. 23, 1994, pp. 487–521. While noting the lack of clear evidence, Ostry agrees that "the internationalization of technology and heightened competition in high-technology industries together seem to be eroding the support of basic and long-run research programs, both private and public." See Sylvia Ostry, "Technology Issues in the International Trading System," OECD, Paris, 26–27 October 1995, p. 16.

[91] Paul David and Dominique Foray, *STI Review,* OECD, Paris, 1995, p. 48. The authors also give the examples of the "reduced ratio of publications of molecular structures whose coordinates are disclosed to total published structures and reduced disclosures of algorithms used in mass spectrographic analysis." Also see Dasgupta and David, "Toward a New Economics of Science." David and Foray also cite S. Hitgartner and S.I. Brandt-Rauf, "Controlling Data and Resources: Access Strategies in Molecular Genetics."

[92] Paul David and Dominique Foray, *STI Review,* pp. 16–17.

A recent Academy report highlights the difficulty of making a policy-relevant distinction between basic and applied research, particularly with respect to federally funded research, while underscoring the importance of public investment in research.[93]

In this environment, the Western system of publicly funded, open scientific endeavor, especially the rapid publication of research, may face serious challenges in the decades to come.[94] This is a cause of profound concern, not least because the current system of open scientific inquiry has been the source of enormous human progress.

The asymmetric development of open and proprietary scientific research efforts may not be sustainable over the long term, in terms of either international interchange or taxpayer support. The exploitation of university research, although usually accomplished with substantial direct and indirect public support, often requires the stimulus of private patents. The appropriate disposition of these patent rights poses complex policy questions which merit further analysis. For example, in some cases publicly funded researchers carrying out work in publicly supported research facilities believe they should maximize the return for their work by licensing it to the highest corporate bidder, whether foreign or domestic. Moreover, at the level of international exchange, without the perception of widespread reciprocal access to equivalent work, pressures to intervene in the free flow of ideas, and ultimately in the functioning of research institutes and universities themselves, are likely to mount, presumably to the eventual detriment of global welfare and scientific advance.

The pressure of global competition is also profoundly transforming industrial research. In the United States, research managers are redirecting resources away from fundamental science and pioneering technology and toward "activities that are more relevant to current product and process development, more likely to produce results that can readily be kept proprietary, and more certain to produce a commercial payoff in the near future."[95] These changes in emphasis, combined with a restructuring of re-

---

[93] *Allocating Federal Funds for Science and Technology,* National Research Council, p. 76.

[94] Dasgupta and David, "Toward a New Economics of Science." See also Paul A. David, David C. Mowery, and W. Edward Steinmueller, "Government-Industry Research Collaborations: Managing Missions in Conflict," paper presented at CEPR/AAAS conference *University Goals, Institutional Mechanisms, and the 'Industrial Transferability' of Research,* Stanford, Calif., 18–20 March 1994.

[95] Richard S. Rosenbloom and William J. Spencer, "The Transformation of Industrial Research," *Issues in Science and Technology, Spring 1996,* pp. 68–74. For a more complete review of the evolution of the U.S. R&D effort, see Richard Rosenbloom and William Spencer, *Engines of Innovation: U.S. Industrial Research at the End of an Era,* Harvard Business School Press, Boston, Mass., 1996. For an industry perspective, see especially chap. 6, "Reinventing Research at IBM" by John A. Armstrong and chap. 7, "Research and Change Management in Xerox" by Mark B. Myers.

search organizations, reflect in part broader changes in innovation systems in the United States and other countries as a result of the end of the Cold War. However, they also reflect the current competitive environment where firms compete in a global marketplace with rivals that do no fundamental research but are quick to exploit the developments made elsewhere. The success of these free-rider strategies and the corresponding reduced ability of traditional innovators to capture returns are reducing incentives for established firms to make significant, and risky, investments in new technologies.[96] This trend may have far-reaching consequences insofar as the innovations brought to the market through industrial research laboratories have been a driving force in the growth of the global economy.

As an initial step, further work on the problems and prospects of international cooperation, particularly in commercially relevant areas such as new materials, information and telecommunications systems, and biotechnology should be undertaken. Further study on the challenges faced by research universities in terms of funding, mission, and relationship to national programs should be undertaken. New modes of cooperative research activities involving both industries and universities should be explored. More fundamentally, an assessment of the extent to which alternative scientific research systems are emerging, and of the potential consequences for the existing system of scientific inquiry and exchange, should be undertaken.

GREATER INTERNATIONAL COOPERATION MAY GENERATE INCREASED FRICTION

The costs, complexity, and risk associated with the development of new technologies provide great opportunities for international cooperation in both the public and private sectors. Common national objectives, and in some cases the scope of these objectives (e.g. assessing global warming), provide strong incentives for cooperation among national programs.[97] In other cases, close public-private cooperation is essential to fully realize the benefits of innovative technologies and systems. This is especially true when critical public service functions, such as civil aviation, are involved, as is the case with the Global Positioning System (GPS).[98] The character of

---

[96] Ibid., pp. 69–71. Rosenbloom and Spencer point out that "there is a logical fallacy in *every* firm planning to be a 'fast second' with the next new technology" (italics added).

[97] For an excellent review of issues associated with large science projects, see Josef Rembser, *Intergovernmental and International Consultations/Agreements and Legal Co-operation Mechanisms in Megascience: Experiences, Aspects, and Ideas,* OECD, Paris, 1995.

[98] For a discussion of the progress and issues associated with the Global Positioning System, see Supplement D.

private innovation is also changing, with the nature and scale of investment now required for the commercialization of fundamentally new technologies such that a single firm is rarely able to dominate.[99] Increasingly, firms with comparable technological capabilities coordinate their activities in order to share costs, risks, and ultimately benefits.[100]

These powerful drivers of cooperation are at the same time a source of greater system friction. As cooperation has become more widespread, addressing more technologies deemed to be of strategic national interest, acute policy differences have emerged among industrial countries. Burden-sharing, equitable technical and technological contributions, and effective and adequate intellectual property protection are all sources of friction. The potential for friction is likely to rise as the importance, scope, and intensity of international interaction increase, unless an international consensus can be reached on appropriate principles and practices for such cooperation.[101]

For example, a number of Japanese technology development programs envisage an important international component. Several large U.S. firms such as Motorola, IBM, United Technologies, General Electric, and the Stanford Research Institute have participated in joint research and development programs designed to pursue Japanese national research objectives.[102] These projects, which range from micromachine technology, to supersonic propulsion systems, to new models for software architecture, offer opportunities for foreign companies to participate in significant development programs. However, the "basic rule" for intellectual property resulting from Japanese government-sponsored research projects is that the intellectual property, including patents and copyrights, is first owned by the Japanese government.[103] Project administration also involves close control by the sponsor-

---

[99] Richard S. Rosenbloom and William J. Spencer, "The Transformation of Industrial Research," pp. 69–70.

[100] Ibid., p. 70. See the section on Strategic Alliances for a discussion of the motivations and types of cooperation among firms.

[101] The issues involved in international cooperation are currently under review at the OECD. The OECD Committee for Scientific and Technological Policy met in September 1995 to review, inter alia, the importance of devising mechanisms to facilitate effective international cooperation both on large-scale science programs and on technology development among national programs and at the enterprise level. See the Ministerial Communique and related discussion papers, DSTI/STP/MIN (95) 1, 2, 3, OECD, Paris, 1995.

[102] Gregory Rutchik, *Japanese Research Projects and Intellectual Property Laws,* p. 8 and appendix 2. This trend toward greater internationalization of Japanese technology programs may be reversing. A number of recently announced programs, in semiconductors for example, are confined to Japanese producers, although this development may be mitigated by the numerous strategic alliances in this sector.

[103] Ibid., p. 31. Work done completely independently of the research project is excluded. However, the Japanese government retains full ownership of copyrights created under directly sponsored research. Similarly, patents resulting from nationally sponsored research projects are subject to the regulations of Japanese patent law. In some cases, lower government ownership provisions can be negotiated.

ing institutions, including regulation of participating researchers, accounting requirements, and "the right of the supervising government agency to inspect the research program."[104] This is not to say that this kind of cooperation cannot be mutually beneficial; it does highlight the importance of different national practices and their potential for friction among both project participants and the home governments of participating companies.

As international cooperation continues to expand, particularly cooperation involving both public and private participants, agreed procedures for dispute settlement are likely to become increasingly important. Explicit mechanisms to resolve disputes between the program participants and the sponsoring government should be established. Moreover, where negotiations are unable to resolve disputes, the wide variation among national laws suggests the need for a clear determination at the outset concerning which law will be applicable.[105]

DIFFERENT MODES OF COOPERATION

The growth in cooperation in the development of new technologies and products can mask significant differences in the way in which that development is organized. The requirements for cooperation and the potential for intergovernment friction are quite different, depending on the form the cooperation takes. At the most basic level, it is important to differentiate among three broad, though conceptually distinct, categories: (1) public-public cooperation among national or supranational authorities; (2) public-private cooperation between public authorities or quasipublic entities on the one hand and private corporations or universities on the other; and (3) cooperation among private entities. The second category, cooperation involving national objectives, public funds, and private entities, is often the most complex. [See the section on eligibility for participation in national programs below.] The third category, involving "purely" private sector alliances, is, however, both the most rapidly growing and probably the least understood.

PRINCIPLES OF INTERNATIONAL COOPERATION

Multilateral institutions have a role to play in developing widely accepted principles for cooperation in the development of new technologies. In this context, cooperative programs such as the Real World Computer Partnership (stressing new applications of opto-electronics technology), the

---

[104] Ibid., p. 34. This pervasive requirement raises intellectual property issues which, while surmountable, necessitate early negotiation among participants.

[105] Ibid., p. 46.

BOX C. DRIVERS OF COOPERATION IN THE SEMICONDUCTOR INDUSTRY[106]

Costs, Risks, and Dispersed Expertise

The semiconductor industry typifies the pressures for increased cooperation. The high costs of product development and of building new factories have led to increased international cooperation in semiconductor product development. Most major semiconductor manufacturers have international joint ventures aimed at producing next-generation memory, logic, and other semiconductor devices. These ventures generally result in factories to produce product and in the sharing of current manufacturing processes.

Standards

Partly as a result of growing private cooperation (see Box E), there is greater international cooperation in the development of standards on semiconductor materials, test methods, microchip interfaces, and other areas where standards can promote industry growth. While international participation in standards development is without doubt beneficial to the semiconductor industry as a whole, how to carry out such cooperation is a major issue which the semiconductor industry and the nations which are leaders in this area must address.[107]

Education and Research

While there is little controversy over international participation in support of basic science and education, there is some concern that the United States is carrying the major cost of research in universities, especially because the technology and graduates flowing from this research investment are often used effectively in other countries which do not maintain comparable research establishments. The training of students is of course a global good, and their U.S. experience and training in the United States may ultimately benefit U.S. suppliers. The growth in research contracts between foreign-based companies and U.S. universities raises complex questions which merit further research.

Environment, Safety, and Health

A growing international consensus on environmental protection has contributed to the belief that cooperation on standards and on environment, safety, and health (ES&H) on an international basis is worthwhile. The global nature of the semiconductor business and the establishment of manufacturing facilities all over the world are powerful incentives for corporations to seek, and governments to encourage, common ES&H standards. For example, the number of countries that are willing to compromise ES&H standards and that are attracting manufacturing sites is decreasing, reflecting the growing global awareness of ES&H issues and the emerging consensus that the elimination of certain chemicals and materials in semiconductor manufacture is a common interest. Similarly, there is increased international interest in standards for "green products" in the computer industry and the ISO 9000 and 14000 production standards.

---

[106] This section draws heavily from William J. Spencer, *SEMATECH,* 20 November 1995 contribution to Steering Committee deliberations. See also the discussion of the new international cooperative initiative on the next-generation wafer standard—the I300I program—in Box D below. See also the discussion of the origins and accomplishments of SEMATECH in Supplement B.

[107] This point was underscored by Samsung's Y.S. Kim, who observed that "in

Intelligent Manufacturing Systems (IMS) Partnership, the Super Sonic Jet Propulsion Project, and the numerous European Union (EU) cooperative programs are of great relevance.[108] The IMS Partnership, for example, provides a framework for international collaboration in advanced manufacturing involving six industrialized regions.[109] These multilateral efforts offer useful lessons regarding the principles and organization of cooperative research.

Work on principles which incorporate the lessons of these programs should continue. Further efforts to improve understanding of differences between national programs should be undertaken. Effective treatment of these questions will require coordination among trade and technology policymakers at both the national and international level. The negotiation now under way at the OECD on a Multilateral Accord on Investment may provide an opportunity to undertake further work on related issues (see below).

CHALLENGES TO COOPERATION

Despite the many potential benefits offered by collaborative research, collaboration can also involve costs and risks. For example, the coordinating mechanisms required for international cooperation may increase total project costs (though these may be offset in part by increased benefits of access to greater expertise). In addition to increased management complexity, there are other obstacles to international cooperation, such as:

• the challenge of identifying cooperative projects of equal interest to all parties;

• the difficulty of distributing costs and benefits in an equitable manner;

• the related problem of ensuring that participants are providing their most advanced technologies and top-level personnel for collaborative projects;

• the transfer of critical scientific and technical knowledge to foreign participants, especially the tacit skills necessary for commercial applications;

• socio-cultural differences and divergent expectations among the partners; and,

• the absence of reliable mechanisms guaranteeing long-term commitment to projects (this is especially true for cooperative efforts involving public authorities).

---

semiconductors...we have come to a point where we have no choice but to cooperate," a perspective supported by Nortel's Claudine Simson and Motorola's Owen Williams in their presentations to the conference *Sources of International Friction and Cooperation in High-Technology Development and Trade, 30–31 May 1995.*

[108] For a review of the issues associated with participation in Japanese technology programs, see Gregory Rutchik, *Japanese Research Projects and Intellectual Property Laws.*

[109] See Coalition for Intelligent Manufacturing Systems and the U.S. Department of Commerce Technology Administration, *Intelligent Manufacturing Systems.*

Lastly, the desire for cooperation, and the increased resources and expertise it can bring, is frequently offset, particularly among bureaucratic elites, by a sense of loss of national leadership, prestige, and control of the cooperative enterprise.[110]

National security considerations can also be an impediment. The risk of the transfer of technologies with proven or potential military applications can raise obstacles both for high-technology firms and for government-to-government military-related research. Even when the partner is an ally, it can be difficult to design a framework that prevents transfer of technologies to third parties or proliferation of technical capabilities.[111] In an age of dual-use technology applications, the United States and other countries are increasingly seeking to integrate the rapid advances in commercial technological capabilities in weapons systems, while benefiting from rapidly declining costs (see below). Separating out the commercial developments from the military applications can be difficult. This adds further complexity to international efforts to cooperate.

NATIONAL AND INTERNATIONAL CONSORTIA

Given the powerful pressures driving public-private cooperation, industrial consortia are likely to become more important (not less) as vehicles for managing the costs and risks associated with developing new technologies, processes, and standards. Though there are significant costs associated with the establishment and management of consortia, returns on investment can be significant.[112] Some analysts find that national consortia have resulted

---

[110] *International Partnerships in Large Science Projects*, Office of Technology Assessment, Congress of the United States, July 1995, p. 101.

[111] Ibid., p. 110, and *Arming Our Allies: Cooperation and Competition in Defense Technology*, Office of Technology Assessment, Congress of the United States, May 1990. The FSX controversy, which concerned the joint production agreement between Japan and the United States for the production of a new-generation fighter aircraft, is a case in point.

[112] See Peter Grindley et al., "SEMATECH and Collaborative Research," p. 725. Citing several recent studies, the authors note that "the hypothesized advantages of collaboration in research include the ability of participating firms to lower costs and spread risks, reduced duplication in their R&D investments, and the exploitation of economies of scale in the R&D process. Research consortia also may be able to appropriate more of the returns to their R&D investments and internalize more of the interfirm spillovers that discourage R&D investment by individual firms..." See B. Bozeman, A. Link, and A. Zardkoohi, "An Economic Analysis of R&D Joint Ventures, *Management and Decision Economics,* 1986; M.L. Katz and J.A. Ordover, "R&D Cooperation and Competition," *Brookings Papers on Economic Activity,* 1990; and David Mowery and Nathan Rosenberg, "New Developments in U.S. Technology Policy: Implications of Competitiveness and International Trade Policy." *California Management Review,* vol. 32, no. 1, Fall 1989.

in the "bundling" of company R&D efforts, with the result that duplicative efforts at the firm level are reduced and program goals accomplished at a lower cost to the industry as a whole.[113] Cooperative efforts can also serve to stimulate international cooperation by providing "easy mechanisms" for identifying technological obstacles best met through international collaboration. Moreover, national consortia are highly effective mechanisms for the diffusion of technological advance derived from collaborative activities and can contribute to the conditions necessary to capitalize on the fruits of such collaboration.[114]

The operation of effective consortia requires, above all, agreement on achievable common goals based on a sense of shared interests. It also requires skilled management and a strong long-term commitment on the part of participants to make available adequate resources in terms of both high-quality personnel and financial support. Senior management of participants must be regularly involved in the strategy of the consortium, with meetings and transparent communication with all levels of member companies.[115]

Given the rapidly rising costs of developing new technologies, the geographic dispersal of expertise, and the need to agree on common standards as a means of assuring market access for final products, international collaboration—including consortia—is likely to become more prevalent (see Box D below). This is, however, less likely to involve the internationalization of existing national consortia than to generate new cooperative mechanisms. Applying the lessons of effective consortium management to these

---

[113] Douglass A. Irwin and Peter J. Klenow, *SEMATECH: Purpose and Performance,* Paper Prepared for the NAS Colloquium on Science, Technology and the Economy, Irvine, Calif. 20–22 October 1995. The authors also point out that the GAO survey of executives from SEMATECH found that members were generally satisfied (though a few founding members left the consortium). They also cite a report stating that "Intel believes it has saved $200 to $300 million from improved yields and greater production efficiencies in return for annual Sematech investments of about $17 million" (pp. 5–6). For a summary review of the nature and accomplishments of the SEMATECH program see Supplement B below.

[114] William J. Spencer, *SEMATECH.* See also Daniel I. Okimoto, *Between MITI and the Market: Japanese Industrial Policy for High-Technology,* Stanford University Press, Stanford, Calif., 1989, p. 67. The author observes that Japanese national research projects, which bring together talent from leading companies and government laboratories, offer an excellent way to leapfrog ahead of the competition. However, Glenn R. Fong argues that the VLSI consortium (and especially its central research facility) proved ineffective in developing a *cooperative* research effort among participating companies. Government subsidies played a crucial role in forcing the pace of technological advance, but program efforts were dominated by individual firms pursuing different technological approaches. See Glenn R. Fong, "State Strength, Industry Structure, and Industrial Policy: American and Japanese Experiences in Microelectronics," *Comparative Politics,* vol. 22, no. 3, April 1990, pp. 290–293.

[115] William J. Spencer, *SEMATECH.*

endeavors will pose a significant but surmountable challenge to the international R&D community.

BEST PRACTICE FOR NATIONAL PROGRAMS OF TECHNOLOGY DEVELOPMENT

The success of national programs to support the development of new technologies and associated industries depends on a variety of conditions, some of which reflect significant historical, cultural, and economic differences among the major trading nations.[116] There are, nonetheless, features common to the most successful national policies. These may include some, but not necessarily all, of the following characteristics:

• **Broad agreement on the need for public-private cooperation** to shoulder the costs and risks of developing new technologies;
• **An effective and diverse organizational structure**. Successful programs are usually backed by extensive resources, including respected government institutions with well-trained, highly motivated staff who see their core mission as the improvement of their nation's competitive position through effective technology policies;[117]
• **Substantial resources** are allocated, with a high degree of predictability over extended periods of time, to develop core technologies, usually through government-industry partnerships and contracts. Usually these arrangements involve multiple, often competing, firms willing to participate in high-risk, high-payoff precompetitive research;
• **Multiple mechanisms** are deployed. The best government-industry relationships are flexible, with different policy instruments and mixes available to address the needs of particular industries and technologies;[118]
• **Project selection is carried out cooperatively with industry.** Projects are developed in pursuit of government missions, or to ensure the develop-

---

[116] Richard R. Nelson (ed.), *National Innovation Systems: A Comparative Study,* Oxford University Press, New York, 1993.

[117] See Alan Wm. Wolff et al., *Conflict among Nations.,* pp. 8–12. The authors observe that "some of the most successful trading partners—such as Japan, Korea, the European Community, and Germany—are characterized by a fragmented trade policy structure and weak horizontal mechanisms for resolving conflicts, yet have generally been able to implement trade and industrial policies that have advanced the national commercial interest. This is attributable in part to the existence within these countries of individual bureaucracies with sufficient strength to advance nationalist economic objectives despite internal opposition, and in part to an underlying consensus, transcending bureaucratic conflicts, that the successful performance of national industries in international competition...is important to the nation and should be fostered." Ibid., pp. 8–9.

[118] See also Sylvia Ostry and Richard Nelson, *Techno-Nationalism and Techno-Globalism,* chap. 3. See also Gregory Tassey, *Technology and Economic Growth.*

ment of new, enabling, or broadly applicable technologies.[119] Public officials work closely with industry on an iterative basis both to identify the appropriate technology focus and to select projects;[120]

• **Projects and programs involve substantial private contributions** to share the cost of the project. The interest and financial support of multiple firms is a key condition for government contributions;

• **Governments rely on private management** of projects and programs. Government plays a major role in the development of new programs and projects, but the final selection process, the management of the projects, and the exploitation of the results are left to industry;[121]

• **Efforts are long-term. This is critical to the success of national technology programs.** As with private sector strategies, programs of national economic development sometimes fail and sometimes succeed. When they succeed, the positive consequences for the national economy (and the negative consequences for competitors) can be dramatic. Moreover, as with the strategies of private corporations, the success or failure of national development strategies is a process of constant learning and innovation, not a one-time event. The sustained commitment of many nations to these programs, even in the face of frequent failure, attests both to their commitment to develop new sources of economic growth and to the competitive benefits these programs can bring their citizens.[122]

PRODUCER VERSUS CONSUMER ECONOMIES: DIFFERENT GOALS

Criticism of targeted support for technology development is unlikely to diminish, regardless of the success or failure of these programs. Indeed, both may elicit criticism. It is important to recognize, however, that this

---

[119] *Allocating Federal Funds for Science and Technology,* p. 21.

[120] Michael Porter believes that Germany has a good record in updating technology because most government-funded research takes the form of joint projects, with research institutes involving firms or incentives for company research. See Michael Porter, *The Competitive Advantage of Nations,* p. 620. See also Jeffrey A. Hart, *Rival Capitalists,* especially chap. 5, which discusses the strengths and limitations of the decentralized German approach to industrial policy. Hart stresses the importance of the Fraunhofer Institutes as a bridge between universities and industry and a provider of alternative channels for the diffusion of information and technology. P. 185.

[121] Several of these points are discussed in *Key Foreign Industrial Competitors: Selecting Core-Technology Products,* U.S. Technology Administration, Department of Commerce, Washington, D.C., October 1992, pp. 2–6. See also the discussion of SEMATECH in Supplement B.

[122] For a discussion of "Leading Indicators of National Competitiveness" see the National Science Board *Science and Engineering Indicators 1993,* chap. 6. One of the indicators is national commitment, that is, "evidence that a nation is taking directed action to achieve technological competitiveness."

criticism, at least in its most articulate form, is concentrated largely in a few Western countries. Other major technologically advanced countries— or those aspiring to become so—benefit from a broad (or broader) societal consensus on the need for a wide range of government support for national efforts to acquire, develop, and disseminate new, enabling technologies.[123]

Criticism of national technology policies may also mask a philosophical rift between participants in the international economy. As noted above, some countries have consciously adopted policies designed to favor growth over consumption, and producers over consumers; they see the acquisition of high-technology industry as a major national goal to be aggressively pursued.

In the pursuit of this high-technology prize, decisionmakers in countries as diverse as Korea, Germany, France, Japan, Singapore, Malaysia, Taiwan, and now China do not accept the premises of Anglo-American economics with its emphasis on the importance of consumer welfare and global allocative efficiency. Decisionmakers concerned with national economic development focus on the role of government in creating the conditions *and providing direction* for the development of the national economy. They do not accept the premise that the market alone, composed of the individual decisions of consumers, would necessarily encourage development where it will do the nation the most good—that is, in terms of national production capacity in leading sectors.[124] Instead, they believe economic development, even by private capital, often requires a coordinated exercise of central power.[125]

---

[123] Richard Samuels, *Rich Nation, Strong Army,* p. 319 and *passim.* Some countries, such as Germany, have long-standing, multifaceted, and often effective technology programs, but nonetheless do not enjoy such consensus. Personal communication to National Research Council from J. Nicholas Ziegler, MIT Sloan School of Management, 6 December 1995. See also Ziegler, "A Capabilities-Based View of German Technology Policy," commissioned by the National Research Council for the project *Friction and Cooperation in High-Technology Development, Competition, and Trade,* 1995. U.S. technology development programs tend to be justified on national security grounds, or for specific missions such as space exploration, health, or energy development, with limited support for economic development programs per se, except at the state level.

[124] James Fallows, *Looking at the Sun,* p. 183. In addition to Friedrich List, this view was, of course, advocated by Alexander Hamilton in his *Report on Manufactures* and reflected policies practiced by England such as the "Act of Navigation," which required that goods going to and from England be carried by English ships—a policy endorsed by Adam Smith in *The Wealth of Nations.*

[125] James Fallows, *Looking at the Sun.* In addition to government direction, institutional factors also play a major role with respect to *private* investment. In the United States, for example, a recent study found that "although U.S. financial markets are highly efficient, there are significant distortions in the allocation of capital within the private sector" that act to reduce business investment, and especially long-term and intangible investments, thereby reducing the pace of innovation in the U.S. economy. These distortions are exacerbated by the failure of federal regulations to keep pace with change in the global economy. For a broader

The focus of these producer-oriented national policies is therefore not on maintaining rule-based competition, either domestically or internationally, but on economic results in terms of the national capacity to generate wealth and therefore national power. Particularly in Asia, the countries that became subservient to the European powers in the eighteenth and nineteenth centuries saw this loss of autonomy as a result of their inability to match the European capability in manufacturing.[126] Consequently, this view of economic relations among states does not correspond with the traditions of Adam Smith and David Ricardo, which view trade as a positive-sum game. It corresponds better with Friedrich List's more zero-sum conception of national economic interactions. In this view, trade is not just a game requiring "a level playing field" for "fair" competition, but is rather a contest in which some nations lose their independence and control of their destiny as a result of their relative economic performance in the community of nations.[127]

The tensions that are often associated with trade disputes in high-technology industries are often exacerbated by the belief, widely held in Anglo-American circles, that countries whose trade practices do not reflect certain (Anglo-American) assumptions are "cheating," that is violating the rules of the trading system and thereby undermining the system itself.[128] Other

---

discussion of these issues, see Robert Denham and Michael Porter, *Lifting All Boats: Increasing the Payoff in the Private Investment of the U.S. Economy*, the Report of the Capital Allocation Sub-Council to the Competitiveness Policy Council, Washington, D.C., September 1995, p. 1 and *passim*.

[126] James Fallows, *Looking at the Sun,* p. 184.

[127] List emphasized the productive power of manufacturers as central to national security, arguing that "war or the very possibility of war makes manufacturing power an indispensable requirement for a nation of the first rank." For a comparison of the views of the early political economists, see Richard Samuels, *Rich Nation, Strong Army*, pp. 4–20. This view has long been the rationale for U.S. defense programs (see Supplement C). For a historical view of the defense contributions to American development, see Geoffrey Perret, *A Country Made by War: From the Revolution to Vietnam—The Story of America's Rise to Power*, Random House, New York, 1989.

[128] James Fallows, *Looking at the Sun,* pp. 189–190. Robert Malpas also addresses the tendency of British electronics companies to complain that their competitors in other countries are subject to different rules or expectations, such as the lower returns on capital required of German and Japanese companies. Malpas observes simply that since these economies are doing well, one must draw the appropriate policy consequences. See Robert Malpas in R. Landau and N. Rosenberg, *The Positive Sum Strategy*, National Academy Press, Washington, D.C., 1986, p. 111. More broadly, Fallows characterizes the difference in perspective between the two systems as follows: "One economic system (consumer-oriented economies) operates as if it does not have to make the largest decisions about national purpose, *except* when the system is being attacked from the outside in time of war. The other (producer-oriented economies) operates as if the state *always* has a role in continuing to guide [the economy]. It is the interaction between these visions, rather than the rightness or wrongness of either of them, that creates problems..." Fallows, *Looking at the Sun,* p. 223.

nations see economics "not as a matter of right or wrong, of cheating or playing fair, but rather merely a matter of being strong or weak."[129] Behind the moral judgments is the presumption that there is only one way to play the game fairly.[130] While some nations seem to be concerned more about fair play and obeying the rules than about results, other nations observe the principle of defending their own interests and seek to ensure results consonant with their national objectives.[131] From this perspective, national decisionmakers may decide to channel capital, usurp patented products or processes, or discriminate against foreign products. These national policies may be shortsighted—but they do not represent violations of a moral code. They are national decisions.

Recognizing the differences in national perspective, and moderating the moralistic tone with which these issues are often discussed, would contribute to reduced international tension and, for some countries, perhaps lead to improved policymaking.

INTERNATIONAL ELIGIBILITY FOR PARTICIPATION IN NATIONAL TECHNOLOGY PROGRAMS[132]

In this competitive environment, international participation in national technology programs will be subject to conflicting pressures. As noted above, there are often powerful public and private incentives for increased international cooperation. A recent OECD analysis notes three separate, but not mutually exclusive, rationales for international cooperation. For example, cooperation in technology development can reduce technical risk and capital costs and also serve as a means of achieving mutually acceptable international standards to facilitate commercialization. A second cat-

---

[129] Ibid.

[130] This is not the case. Views differ within as well as between countries. For example, even in countries that have long-standing programs to support high-technology industry, there is no consensus on the appropriate role for the state. Kantzenbach and Pfister point out "the discrepancy between theory and practice of German economic policy," contrasting the noninterventionist statements of federal economic ministers with the interventionist policies practiced by departments for energy, steel, shipbuilding, and aircraft manufacturers. Kantzenbach and Pfister in Koopman and Scharrer, *The Economics of High-Technology Competition and Cooperation in Global Markets,* p. 273. Similar discrepancies are also found between the policy statements and practices of other nations.

[131] Fallows, *Looking at the Sun,* p. 448.

[132] In policy debates in the United States the terms "foreign eligibility" and "foreign access" are sometimes used interchangeably. The use of the term access implies a level of automaticity that in fact exists for neither American nor foreign-owned companies. A range of criteria is applied to eligible companies. Particularly in collective undertakings (e.g., consortia), there is no guaranteed access even for American companies. A variety of factors apply, as noted in this section.

egory arises from the special interest that smaller or less technologically advanced countries have in accessing the R&D programs of more advanced countries.[133] A third and rather different category is the promotion of international collaboration among R&D institutions (for example within the European Union). All of these factors are likely to grow in importance in the decades ahead. At the same time, nationalistic approaches to technology development are likely to continue to pose obstacles to increased cooperation.[134]

For the technologist, the policy dilemma is that some element of conditionality may be a sine qua non for continued public support for publicly financed programs. Completely unrestricted access to national programs for all potential participants is likely to prove politically fatal to such programs, with potentially serious welfare losses in absolute and relative terms for that nation. On the other hand, candidate companies in possession of technological assets able to meet the requirements of a program will remain attractive to the program managers, regardless of nationality. Similarly, government measures to restrict the global application of the output of cooperative national programs are self-defeating. Companies with worldwide operations are unlikely to participate in overly restrictive programs, and in any event would be hard put to compartmentalize "national" technologies within their global production systems.[135]

International participation is likely to remain constrained by the need to justify the allocation of public funds for the development of technologies by having ownership and exploitation of the technologies and processes

---

[133] Martin Brown, *Impacts of National Technology Programmes,* pp. 41–42. Brown points out there is an implicit trade-off between promoting national competitiveness, and the economies of scale to be gained from participation in international programs. Ibid. For countries with less-developed research infrastructures or for companies in need of cutting-edge technological solutions, the attraction of international cooperation is evident.

[134] The longstanding reluctance of the Japanese defense industry to cooperate *on a reciprocal basis* was described in a recent National Research Council assessment. See *Maximizing U.S. Interests in Science and Technology Relations with Japan: Report of the Defense Task Force,* National Academy Press, Washington, D.C., November 1995. This point is discussed in the section on dual-use technology below.

[135] Efforts by the Department of Energy in 1992–1993 to require U.S. companies participating in cooperative research and development agreements (CRADAs) to apply the resulting innovations primarily to production within the United States created substantial delays in expanding DOE cooperation with high-technology companies. While U.S. production is seen as a fair return for public support to innovation, the realities of global production networks make the applications of such restraints impractical. See also the remarks by Patrick Windham to the *Symposium on International Access to National Technology Programs, 19 January 1995.* For an assessment of the issues associated with CRADAs, see Rose Marie Ham and David Mowery, "Improving Industry-Government Cooperative R&D, *Issues in Science and Technology,* Summer 1995.

developed remain—at least in part—within the boundaries of the national economy. National benefits tests, whether de facto or de jure in application, are likely to continue. These tests are also likely to continue to include conditions such as a national presence, local production, and a local research capability.[136]

Reciprocal access to national programs is likely to remain a de facto policy consideration, especially in the United States. Participating companies, administrators, and politicians are unlikely to welcome national companies based in countries whose national governments seem to systematically deny foreign access to desirable programs, unless the candidate company is seen as bringing major technological assets to the collaboration.

The imposition of other criteria, such as adequate and effective intellectual property protection, investment regimes providing national treatment[137] and effective rights of establishment, as well as reciprocal access to national programs, will continue to be seen by some as necessary conditions for sustainable cooperation. Indeed, some would argue that these criteria for participation can and do serve as implicit norms for participation in cooperative programs. Moreover, these conditions pose little risk for program objectives if sufficient administrative discretion is retained. For others, these government-imposed conditions are a source of international friction and may undermine important principles of the trading system such as national treatment. Whatever the merits of these views, there are some practical constraints to the "carrying capacity" of national technology programs, though it seems unlikely these limits have been exceeded.[138] Still,

---

[136] Ostry notes that there are usually standard, if unofficial, performance conditions applied for participation in EU programs, in contrast to the reciprocity-oriented conditions applied to U.S. programs. See Ostry, "Technology Issues in the International Trading System." p. 12. Despite its long-standing European presence, IBM was initially rebuffed in its efforts to participate in JESSI (Laura Tyson, *Who's Bashing Whom?* p. 150), although on Siemens' insistence IBM was allowed to later join. In contrast, the purchase of ICL by Fujitsu resulted in ICL's expulsion from JESSI and from the European Roundtable Group, which guides ESPRIT. Kende, "Government Support of the European Information Technology Industry," p. 26.

[137] The term "national treatment" reflects the principle that foreign investors, whatever form their investment takes, should be accorded the same treatment as domestic investors. *Multinationals and the National Interest,* Office of Technology Assessment, p. 48.

[138] In this context "carrying capacity" refers to the practice of adding overt requirements for participation of foreign companies in national programs as a means of advancing other policy goals. These include, for the ATP program, considerations concerning the investment and intellectual property regimes of the home countries of multinational companies seeking to participate. These requirements can, in effect, be seen as signals by the Congress concerning appropriate policies for participation in U.S. national programs. See the presentations by Pat Windham and Daniel Price to the *Symposium on International Access to National Technology Programs, 19 January 1995.*

national investment regimes are unlikely to change solely as a result of a national company's being barred from participation in a cooperative project.

These considerations are compounded by other factors which collectively undermine the conditions for successful international cooperation. These include

- asymmetries in the structure and funding of national programs;
- the different technological competencies and assets nations or firms bring to a cooperative enterprise;
- the related perception that some countries are not contributing their "fair share" to basic research; and
- inadequate and ineffective intellectual property protection, and investment regimes which discriminate against foreign acquisition, and fail to provide—formally or informally—national treatment.

The problem of asymmetries in national technology programs may become more acute as advanced high-technology companies in countries with significantly less-developed research infrastructures seek to participate in national programs of the leading industrial countries.

Even in programs formally open to foreign participation there is considerable latitude for administrative discretion with respect to participation by foreign firms. Because rules governing U.S. technology programs tend to be transparent, they are therefore subjected to criticism. In other countries and regions, unofficial criteria for participation are frequently applied.[139] These criteria can range from the perceived technological capabilities of the candidate company, the reputation of its management, and its competitive practices and position, to the candidate company's willingness to agree to performance requirements. All these factors condition both the enthusiasm with which a company's candidacy is received and its prospects for success.

Notwithstanding recent calls for negotiations on access,[140] a formal agreement may prove difficult in light of the underlying differences in scope, rationale, structure, funding, and accessibility of national programs. Less formal international understandings, however, may offer a means of assuring greater transparency, and ultimately greater foreign participation in government-funded civilian research and development programs. However, even informal international commitments may prove ineffective in furthering international cooperation unless there is a sense of shared burdens and equivalent contributions.

While efforts should be made to reduce discriminatory restrictions on foreign participation in national technology programs, formal declarations

---

[139] Ostry, "Technology Issues in the International Trading System."

[140] *Trans-Atlantic Business Dialogue: Overall Conclusions,* 11 November 1995, Seville, Spain, recommendation III-9, p. III-3.

are not likely to be effective in advancing international cooperation. **More usefully, a sustained effort to reduce conditionality, perhaps through the construction of an objective, internationally accepted national benefits test, might be undertaken on a multilateral basis.**[141] A multilaterally agreed national benefits test might take into account conditions such as the level of a company's R&D presence in the country funding the technology program, specific technological and financial contributions the company would undertake to make to the project, and agreement by participating companies to conduct the R&D funded under the project in the sponsoring country and to manufacture some agreed-on portion of the products derived from the research within the host country.[142] Many of these conditions are already applied on a de facto basis in some jurisdictions. The advantage of a multilateral effort would be to make these requirements more transparent, establish agreed-upon guidelines, and focus on the contributions to the national technology base rather than on corporate "nationality."

In addition to the considerations noted above, international cooperation will continue to be conditioned by the degree of agreement on shared priorities, equitable technical contributions (not merely financial contributions), and a shared capacity to exploit the results of cooperation. These and other factors, such as the contestability of end-product markets, are fundamental elements of sustainable cooperation.

One avenue to improved international cooperation is better information and more transparency. As a first step, an appropriate multilateral organization, such as the OECD, should gather improved data concerning formal rules for participation in national or regional technology programs, supplemented by objective assessments of current administrative practices—i.e., actual foreign participation and its rationale—rather than theoretical "openness." A better understanding of the rules, current practice, and in some cases the absence of rules would be helpful.[143] Ultimately, the basis for sustainable international cooperation is likely to be derived from the combination of support by private sector participants within the host country, the technical or financial needs of the program or agency mission, and a sense

---

[141] This policy option and its rationale are outlined in greater detail in *Multinationals and the U.S. Technology Base,* Office of Technology Assessment, Congress of the United States, September 1994, pp. 33–34. The list of criteria advanced by the OTA authors is more elaborate and potentially more restrictive than the criteria summarized here.

[142] Ibid.

[143] Ostry, "Technology Issues in the International Trading System," p. 12. As noted, the author correctly underscores the lack of complete information on either rules or actual participation by foreign companies in national programs and cites, as an example, the case of the European Commission Framework Program, where "there are no formal guidelines for participation in projects." Membership is negotiated on case-by-case basis or bilaterally, with "unofficial" conditions, including performance requirements, usually a standard feature.

of fair and equitable contribution. When the overriding national goal is autonomous national capability, opportunities for long-term foreign participants will necessarily remain limited. A sustained multilateral effort could also seek to improve understanding of differences among national technology development programs. For example, it could gather improved data concerning formal rules for participation in national or regional technology programs, supplemented by objective assessments of current administrative practices, i.e., actual foreign participation and its rationale, rather than theoretical "openness."

A CASE-BY-CASE APPROACH

Despite these caveats, there are substantial benefits to be gained from increased international cooperation. Their realization is most likely to occur on a case-by-case basis, directly related to other factors such as program goals and investment regimes. Private motivations will remain powerful drivers. An important current example is the semiconductor industry, which now offers a significant opportunity for a major cooperative initiative as the industry moves to the 300mm wafer for semiconductor devices. This is a major technological challenge, which many in the industry believe will best be met collectively. The American semiconductor consortium SEMATECH has played an instrumental role in establishing an international consortium to meet this challenge. (See Box D.)

The desire of otherwise fierce competitors, such as the major semiconductor firms, to cooperate on a common standard underscores the technical and financial challenges the industry is facing in moving to a new standard. Industry-initiated efforts at cooperation are most likely to succeed when backed by a supportive policy framework. In some cases, public initiatives can provide a critical catalyst for cooperative efforts to achieve goals identified by industry.[144] Efforts initiated by public officials less attuned to the complexities of a given industry may experience corresponding difficulty setting and achieving proper objectives.[145] Cooperation offers the greatest

---

[144] See Richard Brody, *Effective Partnering*, Office of Technology Policy. Programs involving partnering among multiple companies can serve to remove barriers to public-private collaboration, but also to collaboration between firms. "Companies report that...just applying for federal programs" can be immensely valuable regardless of whether the firms actually receive funding. Pp. 53–54. For a similar view, see European Commission *European Report on Science and Technology Indicators, 1994.*

[145] The EUREKA initiative, launched in 1985 by France, was intended to provide an alternative to more bureaucratic European Community programs. EUREKA is designed to be more bottom-up and driven by the interest of firms; although usually encouraged by the prospect of public support, EUREKA has no independent funding. The Framework program mobilizes about 5 billion ECU per year; EUREKA mobilizes up to 2 billion ECU per year. See the

BOX D. A REAL-WORLD CASE:
INTERNATIONAL COOPERATION ON THE 300MM WAFER

The growth and increasing productivity of the semiconductor industry are fueled partly by increasing wafer size. The silicon wafer is the platform on which microchips are built. The standard wafer size in 1970 was about 30mm. Today the standard in modern semiconductor manufacturing facilities is 200mm. The next generation wafer will be 300mm, and it is expected to become the dominant platform for manufacture early in the 21st century. The conversion to 100mm wafers in the mid-1980's led to an interesting situation in the United States. While there was some work on international standards, most of the world converted to 100mm wafers while the United States in some instances converted to four inches, i.e., 101.6mm. This meant that fixtures and tools could not handle both sizes of wafers. It is important that this error not recur in the transition to the 300mm wafer. Agreement on an international standard is important because making and using larger wafers necessitates change in a whole range of technical characteristics of the materials and manufacturing methods.

The cost of changing from 200mm to 300mm wafers is expected to exceed $10 billion, with the highest estimates at $30 billion. By comparison, the last conversion from 6 inch to 8 inch wafers was believed to have cost on the order of $2 billion and was managed largely by IBM, which then made the new technology widely available. Today, however, Intel—the industry giant—is a $14 billion company unable to assume these costs. Moreover, the fabrication facilities which will run the 300mm wafer will be dedicated facilities, unable to accommodate the older wafer size. These facilities are likely to cost on the order of one to one-and-a-half billion dollars each. The standardizing of wafer size is therefore driven in part by the need to avoid imposing customization costs on the tool-making industry.

This dramatic cost escalation has given birth to cooperative efforts in both Japan and the United States. The Japanese effort is principally a national program focused on the Japanese equipment industry, with substantial public funding. The 300mm program launched by a SEMATECH subsidiary is completely funded by industry and open to participants from other nations. The consortium's funding is to be based on equal contributions from participating companies, which now include members from Europe, Korea, and Taiwan as well as the United States.

Though a separate entity, the new consortium will have the major advantage of being able to use SEMATECH facilities. The program will focus on international standards, on setting international requirements for 300mm wafers at 0.25 and 0.18 μ geometries (256 Mb DRAM and 1 Gb DRAM), and on the evaluation of critical equipment for 300mm manufacture. This cooperative program is expected to last for 18-24 months, at which point an evaluation will be made to determine whether it should continue for the development of additional technology related to 300mm wafers.

prospect of success when it is built on specific shared objectives and a clear understanding of the costs and benefits.

STRATEGIC ALLIANCES

Notwithstanding the intense global competition among companies for market share and new, innovative processes and products, the last decade has seen a rapid expansion in another form of cooperation: international strategic alliances. The growth and importance of these alliances may in time lead to a reshaping of the way in which companies cooperate and compete.[146] The causes of this phenomenal increase in corporate alliances are diverse and powerful. They include the evolution of modern technology and productive processes, the need to recover rapidly rising costs in as large a market as possible, and efforts by companies to respond or adjust to government initiatives—on the environment, for example—or to overcome exclusionary government practices.

Technology acts as a major driver of alliance formation. The diversity, complexity, and cost of new technologies are all major sources of the dramatic increases in alliances. Alliance formation is also driven by the need to produce innovative products in an ever-shorter timeframe and to access not only markets and technologies but also the tacit knowledge to deploy them effectively. Technology convergence also plays a growing role. Firms must increasingly manage a diverse array of new technologies, sometimes outside their core competencies. To do so, they seek alliances to aid in managing technology convergence across formerly separate industries. Each of these factors encourages alliance formation as a means of accessing expertise and hedging against technological risk.

Governments also play a major, often decisive, role in driving alliance activity. Government trade, investment, procurement, and regulatory policies present obstacles and opportunities both of which encourage alliances. Restrictions on market access act as incentives for firms to form alliances

---

presentation by Reinhard Loosch, "EUREKA and the Framework Program" in C. Wessner (ed.), *Sources of International Friction and Cooperation in High-Technology Development and Trade.* The Hanover Declaration, the basic charter for EUREKA, states that "the fundamental aim of EUREKA is to raise the productivity and competitiveness of Europe's industries and national economies on the world market through products, processes, and services which have a worldwide market potential and are based on advanced technologies." Ibid., p. 265.

[146] This section draws heavily from the presentations of Carol V. Evans of Georgetown University and Charles White of Motorola to the National Research Council conference *Sources of International Friction and Cooperation in High-Technology Development and Trade, 30–31 May, 1995.*

as a means of bypassing border restrictions.[147] For instance, some governments restrict access to their markets unless would-be exporters supply
critical technologies, manufacturing capabilities, and/or distribution rights
to domestic producers. Governments also sometimes intervene directly in
shaping or encouraging alliances in strategic sectors, such as telecommunications, aerospace, and semiconductors.[148] These interventionist practices
raise important policy issues for countries with relatively open domestic
markets. In the case of the United States, some industries have benefited
from new technologies, processes, and managerial methods learned through
strategic alliances. But for the United States and other technology leaders,
asymmetrical technology flows pose a long-term challenge. Competitive
positioning by firms seeking to monitor or assess the activities of rivals or
to learn about new product lines also motivates alliance formation. Significantly, the scope and intensity of alliance activity are unevenly distributed,
with the highest growth occurring in such high-technology sectors as aerospace, biotechnology, information systems, and the automotive industry.
(Box E reviews the different types of alliance activity.)

These trends in alliance formation and their implications are not well
understood by policymakers. In some cases, alliances represent innovative
efforts to meet technological challenges. In other cases, they are second-
best strategies adopted by multinationals to counteract either discriminatory
policies of foreign governments or actions by publicly controlled firms. In
some special cases, e.g., Airbus, they are the direct result of government
action. The growth in corporate alliances paradoxically has the potential to
create friction at both the national and international level. Some strategic
alliances, involving taxpayer-supported technologies, may generate friction
between domestic companies and national governments. Alliances between
inherently unequal partners can result in the transfer of key technologies to
foreign partners in exchange for short-term gains.[149] In other cases, compa-

---

[147] *Multinationals and the National Interest: Playing by Different Rules,* Office of Technology Assessment, Congress of the United States, Washington, D.C., 1993, p. 116. The issue
of compulsory technology transfers and their impact on the manufacturing base of the United
States and other industrial countries is taken up below.

[148] Alcatel in France and Airbus in Europe are two cases in point. See the discussion of
national champions and sectoral strategies (above) and the encouragement of alliance activity
offered by the semiconductor agreement (below).

[149] Carol V. Evans' presentation to the conference *Sources of International Friction and
Cooperation in High-Technology Development and Trade.* See also *Multinationals in the
National Interest,* chap. 5. This can also occur with domestic partners, although the consequences for national technology capabilities of such transfers are presumably less significant
for technologies transferred domestically than for technologies transferred and developed offshore. For a discussion of the objectives and assets of entrepreneurial companies and large
multinational partners, see *U.S.-Japan Strategic Alliances in the Semiconductor Industry: Technology
Transfer, Competition, and Public Policy,* National Research Council, National Academy Press,
Washington, D.C., 1995

BOX E. TYPES OF ALLIANCE ACTIVITY

R&D Alliances

1. **Licensing agreement**: legal permission to utilize patents or proprietary technology for an up-front fee and/or royalties.

2. **Cross-licensing agreement**: two or more companies give legal permission to use each other's patents or proprietary technology.

3. **Technology exchange**: a swap of proprietary technologies, which may or may not involve a transfer of money.

4. **Visitation and research participation**: the dispatch of researchers to visit, observe, and participate in R&D activities of partner firms.

5. **Personal exchange**: an ongoing and reciprocal program in which researchers from one company spend time working at the partner company.

6. **Joint development**: two or more companies joining forces to develop new products or technology.

7. **Technology acquisition investments**: foreign investments in companies aimed at gaining access to technology, especially in small, start-up or innovative, medium-size firms.

Manufacturing Alliances

8. **Original equipment manufacturing (OEM)**: manufacturing a product for another company, which sticks its label on it and handles all aspects of business activities, including marketing and servicing, as if it had manufactured the product itself.

9. **Second sourcing**: an arrangement whereby a company is given permission to manufacture a product designed and developed by another company as a second source of supply for customers, using the same qualifications.

10. **Fabrication agreement**: use of another company's fabrication facilities to manufacture a product (because the partner either lacks its own manufacturing facilities or wishes to subcontract out the task of fabrication).

11. **Assembly and testing agreement**: components and parts manufactured elsewhere are sent to another company where they are assembled and tested.

Marketing and Service Alliances

12. **Procurement agreement**: a commitment to purchase certain quantities of specific goods or services over a specified period of time.

13. **Sales agency agreement**: exclusive or nonexclusive rights to sell the partner's original products, or products to which value is added, in specified markets.

14. **Servicing contracts**: the provision of follow-up service in foreign markets (often tied to marketing arrangements).

General-Purpose Tie-ups

15. **Standards coordination**: an agreement on common or compatible technical standards, linking devices, systems, and users of different machines.

16. **Joint venture**: two or more firms jointly form a company to develop, manufacture, or market new products.[150]

---

[150] Prepared by the NRC working group for the report *U.S.-Japan Strategic Alliances in the Semiconductor Industry*, p. 10.

nies may barter critical technologies developed at government expense, for access to markets, other technologies, or capital. In some countries the costs, risks, and benefits are calculated by the firm; in others, the calculation is made in consultation with national authorities.

The growth in strategic alliances is changing the traditional terms of international competition. Instead of national champions, international coalitions of diverse national origin may compete for global market share. Similarly, the growth in cross-equity investment and shared production facilities across different countries poses challenges to national technology policy, just as alliances among market leaders, particularly around technical standards, challenge traditional competition policies. These areas require both further analysis and flexible policy responses. Government intervention into the formation of private alliances can significantly affect outcomes, despite the limitations (which vary among countries) on government influence. **While the need for government intervention in private alliance activity is no doubt limited, it is important to recognize both that government actions often generate alliances and that these alliances can have a significant impact on the competitive environment.**

GLOBALIZATION OF R&D?

Reflecting the growth in strategic alliances and direct investment, private R&D activity has increasingly taken on global dimensions. Foreign firms have invested heavily in R&D facilities in the United States and, increasingly, U.S. firms are conducting R&D overseas. U.S. companies have increased their total overseas R&D spending substantially (from $5.2 billion in 1987 to $9.8 billion in 1993).[151] Moreover, the U.S. R&D effort is becoming more geographically dispersed. While half of U.S. foreign-based R&D spending is located in Germany, the United Kingdom, Canada, France, and Japan, substantial increases in spending have occurred in countries such as Singapore, Brazil, Mexico, and Hong Kong.[152]

The United States is also benefiting from these trends. R&D expenditures by foreign-owned companies in the United States have more than doubled, from $6.5 billion in 1987 to $14.6 billion in 1993.[153] Foreign expenditure now accounts for more than 15 percent of total U.S. private R&D, and the rate of R&D spending by foreign-owned companies in the United States is increasing more rapidly than R&D spending by U.S. firms.

---

[151] *Globalizing Industrial Research and Development,* Office of Technology Policy, U.S. Department of Commerce, October 1995, p. 8.

[152] Ibid., p. 29. Hong Kong, of course, will become part of the People's Republic of China in 1997, presumably strengthening the Chinese technology base.

[153] Ibid.

This spending reflects the significant R&D presence of foreign companies, which now own more than 645 R&D facilities in the United States. Japan owns 224, Germany 95, and France 52, with new entrants such as Korea having doubled their facilities in the last three years. Together these foreign-owned companies employ more than 105,000 R&D workers.[154]

To some extent, motivations for overseas investment tend to be similar across companies and sectors. For example, in the electronics sector, U.S. R&D in Japan and Japanese R&D expenditures in the United States were designed to meet the needs of foreign affiliates; monitor technology developments in the foreign market; and assess, acquire, or generate new technologies.[155] More broadly, R&D tends to be the last aspect of corporate activity to move overseas, though foreign production capabilities often result in selective R&D decentralization. Historically, firms move R&D abroad to

- acquire foreign technology,
- customize products for local markets,
- monitor foreign technological developments, and
- gain access to foreign R&D resources, such as universities, public and private research facilities, and highly trained scientists and engineers.[156]

The establishment of foreign R&D facilities can also facilitate the adaptation of products to local product standards and regulations and can result in substantial cost efficiencies. These establishments involve significant benefits to the economy in which they are located through employment, funding of academic research, equipment purchases, and contributions to the national technology base. For example, Japanese-funded R&D in the United States, which increased from $307 million in 1987 to $1.8 billion in 1993, represents a significant addition to U.S.-based R&D.[157]

Notwithstanding the importance of these trends, corporate research and development activity remains nationally based. In the case of the United States, for example, the significant expansion of foreign R&D expenditures can be attributed in part to major acquisitions by foreign multinationals of

---

[154] Ibid.

[155] Ibid.

[156] *Multinationals and the U.S. Technology Base,* Office of Technology Assessment, p. 76. It is worth noting that the United States benefits from the presence of self-selecting immigrants, trained in the U.S. university system, many of whom remain after completing their education. The United States therefore benefits from a net inflow of talent from other countries. Other countries, which have much more restrictive immigration policies, do not, although they also concentrate limited educational resources on their own nationals.

[157] *Globalizing Industrial Research and Development,* Office of Technology Policy, Department of Commerce, p. 29.

research-intensive U.S. firms.[158] Moreover, even for companies with extensive international operations and investments, core technology development remains largely centralized at company laboratories in the home country.[159]

The globalization of R&D therefore lags substantially behind the globalization of production, sourcing, and other business activities. To some degree, this reflects the "normal" evolution of activity from trade to direct investment in production activities, which eventually require R&D support. It may also be explained in part by the fact that, in some industries, production facilities can be established and moved relatively quickly in response to changing market conditions. By comparison, R&D facilities have long lead times and high fixed costs and are difficult to move. This encourages the centralization of basic research and product development.[160] It also reflects management's perception that maintaining the company's core technology competency is a task properly carried out in the corporation's home country.[161]

TECHNOLOGY COOPERATION AND
AN OPEN MULTILATERAL TRADING SYSTEM

Greater international cooperation in technology development is facilitated by an open, market-driven trading system. Long-term cooperative efforts, and the cooperative spirit they presuppose, coexist with difficulty in an environment marked by trade disputes or inadequate respect for the explicit and implicit rules of the game. Consequently, a

---

[158] Ibid., pp. 10–11. The Office of Technology Policy report notes that in addition to the late 1980s surge in acquisitions of companies in industries such as computers, semiconductors, steel and tires, the largest impact on R&D funding was derived from the acquisition of U.S. pharmaceutical and biotechnology firms with large R&D budgets.

[159] *Multinationals and the U.S. Technology Base,* Office of Technology Assessment, chap. 4.

[160] The OTA observation that leading-edge R&D for core technologies is performed at the central labs of the corporate home country is supported by empirical research, which shows that most of the patents of large multinationals are filed in the home country. See Pari Patel and Keith Pavitt, "Large Firms in the Production of the World's Technology: An Important Case of Non-Globalization," *Journal of International Business Studies,* First Quarter, 1991. This phenomenon is also supported by R&D expenditure data by U.S. firms which show that about 90 percent of R&D expenditures by U.S. companies occur at their facilities in the United States. See *Globalizing Industrial Research and Development,* Office of Technology Policy, p. 31. See also J.A. Cantwell and C. Hodson, "Global R&D and British Competitiveness," in M.C. Casson, (ed.), *Global Research Strategy and International Competitiveness,* Oxford, Basil Blackwell, 1991, cited in *Globalizing Industrial Research.*

[161] The analysis advanced by Michael Porter in *The Competitive Advantage of Nations* supports this perception. He argues that "competitive advantage is created and sustained through a highly localized process." Furthermore, he stresses that as a result of the globalization of competition, the role of the home nation is more rather than less important. The multinational's home base is the nation where a firm's strategy is set and the core product and process technology are created and maintained. P. 19.

key condition for sustained international cooperation in the development of new technologies is improved adherence to the principles of a liberal trade regime. Closed national markets, whether through quotas, discriminatory standards, or biased public procurement, undermine the political and policy conditions necessary for effective international cooperation. Reciprocal access to national technology development programs fundamentally requires equal access to end-use markets.

Efforts to further technological cooperation, particularly public/private cooperation, therefore imply parallel efforts to further trade liberalization in areas "within the borders," such as government procurement, national treatment for foreign investment, and effective competition policy. Sustainable cooperation implies a competitive, transparent procurement regime; the right of establishment for foreign investors, including roughly comparable regimes for the acquisition of existing firms; and market access for final products resulting from such cooperation.

STRENGTHENING INSTITUTIONS TO INTEGRATE TRADE AND TECHNOLOGY POLICIES

There are powerful, reciprocal relationships between trade and technology policies. However, the degree to which national policymaking reflects this reciprocal relationship varies a great deal among countries. The coordination of trade and technology policies, with their far-reaching economic and political ramifications, is always difficult to accomplish, even in countries with an appropriate institutional structure. The absence of such a structure makes the process of effective policy coordination especially difficult. These institutional issues are especially relevant with respect to the United States, both because of the impact of U.S. policymaking on the international system and because government restructuring is currently on the U.S. domestic political agenda.[162]

The need for structural reform of U.S. policymaking has been the topic of a growing number of studies recommending a restructuring of the U.S. international economic policy apparatus.[163] The fragmentation of authority

---

[162] See, for example, Alan Wm. Wolff et al., *Conflict among Nations,* chap. 9. See also the article by Paula Stern, "Reorganizing Government for Economic Growth and Efficiency" in *Issues in Science and Technology,* Summer 1996, pp. 67–72.

[163] For one of the most comprehensive reviews of the trade policy background, policy process, and trade strategies of Japan, Germany, South Korea, Taiwan, Brazil and the European Community, as well as the United States, see Alan Wm. Wolff et al., *Conflict Among Nations.* For broadly similar views of the reforms required for the U.S. system, see John J. Murphy and Paula Stern, *A Trade Policy for a More Competitive America,* Report of the Trade Policy Subcouncil to the Competitiveness Policy Council, March 1993, and Paula Stern, *Getting the Boxes Right: New Blueprints for U.S. Economic Policymaking,* Economic Strategy Institute, Washington, D.C., 1995.

which now exists tends to encourage inefficiencies, not least for its "...undesirable separation between policy development and implementation."[164] These structural deficiencies have direct effects on U.S. policy. For example, a recent study noted the tendency of the U.S. government to rely on trade policy as a means of responding to broader issues of U.S. competitiveness, a tendency compounded by a fragmented system of policy development.[165]

Effective policymaking and its execution require appropriate institutions, which take time to build but can have important long-term effects.[166] As competition for high-technology industries becomes more acute, new institutions are needed to better link technology and related economic policies with trade policy formulation and negotiations and with export promotion and control. It is especially important that they have the capacity to assess, coordinate, and implement the various policies impacting the development of national high-technology industries.[167] An integrated approach requires institutions designed to support national capabilities and national firms, while at the same time preserving market-based competition and strengthening international disciplines.

---

[164] Paula Stern, "Reorganizing Government for Economic Growth and Efficiency," p. 68.

[165] See Council on Competitiveness, *Roadmap for Results: Trade Policy, Technology and American Competitiveness.* Washington, D.C., 1993, pp. 5–11. The report observes that U.S. "trade policy has suffered from a lack of effective information exchange between the private sector and government and among the various branches of government, particularly the agencies handling technology and trade policy," pp. 10–11.

[166] Traditional economic analysis often understates the importance of institutions in creating "comparative advantage." This point is made more broadly by Douglass North in "Economic Performance Through Time," his Nobel Prize acceptance speech, December 1993, as reprinted in *The American Economic Review,* June 1994, p. 359. North draws the distinction between economic analysis of how markets function at a point in time and the corresponding analysis of *how economies develop over time*: "There is no mystery why the field of development has failed to develop during the five decades since the end of World War II. Neoclassical theory is simply an inappropriate tool to analyze and prescribe policies that will induce development. It is concerned with the operation of markets, not how markets develop." He notes that "theory in the pristine form....gave it mathematical elegance [and] modeled a frictionless and static world..." However, "when applied to economic history and development it focused on technological development and more recently human-capital investment but ignored the incentive structure embodied in the institutions that determined the extent of societal investment in those factors. *In the analysis of economic performance through time it contained two erroneous assumptions: (1) that institutions do not matter and (2) that time does not matter.* "(Italics added.)

[167] See Alan Wm. Wolff et al., *Conflict among Nations,* p. 12, p. 536, and chap. 9. As an example, the author contrasts the information, analysis, and policy tools available in support of U.S. agriculture with the absence of comparable institutions and resources for U.S. manufactured products. From a comparative perspective, Stern notes that "the foreign governments with which Washington negotiates usually combine the functions of trade negotiation, promotion, policy formulation, and compliance investigation within single agencies... If anything, their less fragmented bureaucracies give them an advantage in pursuing their national economic interests." Paula Stern, "Reorganizing Government for Economic Growth and Efficiency."

In the absence of an integrated approach to international competitiveness in high-technology industries, U.S. policy can impose costs on its consumers and producers while putting unnecessary stress on the international trading regime. In the absence of a strategic vision of high-technology competition, U.S. industry is likely to face competitors supported by a panoply of promotional policies, often including the advantages afforded by a protected home market. In the global competition for high-technology industry, this situation is not a recipe for success, nor a means of maintaining long-term support for the multilateral trading system.

In the absence of effective foresight or alternative policy mechanisms, the United States has traditionally responded, in extremis, to challenges to strategic industries with trade measures. Such measures, absent a coherent policy framework, can impose costs on U.S. producers dependent on foreign imports as well as on the consumers of the final product.[168] Yet trade policy measures tend to be selected because they are often the only instrument available to policymakers.[169] Trade measures also have the added advantage of being off-budget. Consumers, not the government, underwrite the costs.[170] Improving the coordination of technology and trade policy through institutions with the necessary resources and analytical capacity offers a means to avoid unnecessary friction, while maintaining a clear understanding of the stakes for both the national economy and the international system.[171]

More effective national policymaking must be complemented by effective international institutions. Because many of the policy questions associated with the promotion and protection of national high-technology industry will have to be addressed on a multilateral basis, international institutions are likely to play an expanded role. It is important that the relevant multilateral institutions adapt their practices and structures to enable them to effectively engage the international community on these questions. As a first step, improved data collection and better understanding of the nature of international competition in high-technology industries would be a valuable contribution to the international dialogue.

---

[168] The case of dumping duties on displays imported by U.S. manufacturers is frequently cited. See Council on Competitiveness, *Roadmap for Results,* chap. 1. See also Jeffrey A. Hart, "Anti-Dumping Petition of the Advanced Display Manufacturers of America: Origins and Consequences" paper delivered at the Annual Meeting of the International Studies Association, Atlanta, Ga., 1–4 April 1992.

[169] Council on Competitiveness, *Roadmap for Results,* p. 7. The report also concludes that a disproportionate responsibility for addressing U.S. competitiveness has fallen on trade policy, with insufficient attention accorded to technology policy. To some extent, the semiconductor case is an exception in that *both* trade and technology issues were addressed. (See Supplements A and B.)

[170] Laura Tyson, *Who's Bashing Whom?* p. 289.

[171] Alan Wm. Wolff et al., *Conflict among Nations* and Paula Stern, "Reorganizing Government for Economic Growth and Efficiency" make similar assessments.

BOX F. COMPARATIVE ADVANTAGE AND HIGH-TECHNOLOGY COMPETITION

Competition for high-technology industries is quite different from the static, textbook competition between countries with different factor endowments, competing on the basis of "natural" comparative advantage. Competition for these industries is inextricably linked to and affected by government policies. In many countries, policymakers recognize that their objective is not just to profit from the current portfolio of national advantages, but to create these advantages in the first place and to upgrade them over time. In short, comparative advantage in high-technologies is often created by conscious national effort.[172]

The semiconductor industry, among others (see Box D), demonstrates the importance of supportive government policies and their effective exploitation by a dynamic market-oriented private sector. The semiconductor industry "wherever it has developed, has been an explicit target of industrial policy—whether a result of military policy of the United States, or the objective of commercial policy elsewhere in the world."[173] An outstanding example of successful government intervention for commercial objectives in semiconductors is the Very Large Scale Integrated (VLSI) Project, initiated in the mid-1970s by MITI and a number of major Japanese companies. Focused on DRAM development and manufacture, the project recognized the importance of complementary metal oxide semiconductor (CMOS) technology and solid state memory as technology drivers of the entire industry. As noted above, in exchange for a modest investment in this joint government-industry project (some $300 million over four years), the Japanese producers moved from being relatively small players in the global semiconductor business to become the dominant producers of DRAMs by the mid-1980s. The Japanese semiconductor equipment industry also grew dramatically as a result of this investment. Moreover, companies such as Nikon were induced by the government to enter the semiconductor equipment business and today dominate photolithography—the most expensive and most critical manufacturing technology.[174]

---

[172] This is especially true of high-technology industries. Leading examples are the semiconductor and aerospace sectors, both of which have benefited from the highly visible hand of government intervention. For a comprehensive review of national policies to create comparative advantage in semiconductors and related industries, see Thomas Howell et al., *Creating Advantage*. See also Kenneth Flamm, *Mismanaged Trade?* chap. 2 and 4. For aerospace, government support through infrastructure testing facilities and R&D support played a major role in developing the industry, in the United States and elsewhere. See Mowery and Rosenberg, "The Commercial Aircraft Industry," in Richard Nelson (ed.), *Government and Technical Progress: A Cross Industry Analysis,* Pergamon, New York, 1982. See also Box H below.

[173] Laura Tyson, *Who's Bashing Whom?* p. 85.

[174] For an excellent review of the accomplishments of the VSLI project, see J. Sigurdson, *Industry and State Partnership in Japan: The Very Large Scale Integrated Circuits (VLSI) Project,* Discussion Paper No. 168, Lund, Sweden Research Policy Institute, 1986, pp. 121–122. Flamm also provides an overview of the VSLI program. See Kenneth Flamm, *Mismanaged Trade?* pp. 94–113. See also Spencer, *SEMATECH,* p. 3. Spencer also points out,

Supportive government policies, however, are nonetheless not sufficient for international competitiveness. The ability to participate—that is, to compete—effectively in the world market is of fundamental importance. Moreover, this competition is often highly dynamic. Competition, particularly in industries which are knowledge intensive, involves close races among firms, with outcomes depending on human endeavor and continuous learning more than natural endowments.[175] Increases in market share for firms (or national industry) translate into increased volume and experience, which in turn translates into increased advantage.[176] Short-term sacrifice to build market share can lead to reversals in market position; firms can come from behind to capture a leadership position in enabling technologies and high-revenue industries. Market position and cost advantages at any given time thus reflect strategies and the skill with which they are implemented, not just a pre-ordained natural order of comparative advantages.[177]

---

however, that poorly conceived government-industry programs can drain critical manpower into nonproductive programs, citing the U.S. Department of Defense program on Very High Speed Silicon Integrated Circuits (VHSIC) in the mid-1980s. Jay Stowsky supports this point, arguing that military specifications and security requirements impede the diffusion of potentially valuable technologies. See Borrus et al., *The Highest Stakes,* chap. 4, "From Spin-Off to Spin-On: Redefining the Military's Role in American Technology Development."

[175] While not the focus of this analysis, the human factor cannot be understated. In the case of Korea, for example, the availability of well-trained engineers—a result of state policy— and the preference of Korean management for engineers over administrators, coupled with a tight check on overhead, limited middle management, and well-educated labor, have been key elements in the success of the chaebol, the diversified business groups which have led much of Korea's development. Amsden, *Asia's Next Giant,* pp. 9–10.

[176] Scott, *Economic Strategies of Nations,* p. 28.

[177] Ibid. See also Borrus et al., *The Highest Stakes,* chap. 1 and pp. 179–184.

System Integration and System Friction: New Challenges in Trade Policy

The creation and nurturing of high-technology industries presents a special case for policymakers in several respects.

• First, for the reasons outlined above (see Box A), policymakers in industrial and newly industrializing countries are persuaded of the benefits of encouraging the development of high-technology industries within national borders. The drive to acquire competency in enabling technologies and to diffuse this knowledge throughout the national economy is an objective shared by policymakers throughout the industrial world.

• Second, many of the practices inherent in national innovation systems are deeply rooted in national economic systems, and reflect long-standing government support for programs and practices. For many years, these policies were considered to be "within the border," that is, within the jurisdiction of domestic policies and not subject to international review. Indeed, the inclusion of what were formerly considered to be domestic policies in the GATT negotiations was a major source of contention in the Uruguay Round.[178]

During the 1980s, national and regional policies to encourage high-technology industry played a major role in the increased international trade conflict centered on high-technology systems such as aircraft, superconductors, and telecommunications products; technology-intensive products such as automobiles and steel; and key components such as semiconductors and machine tools. The Airbus dispute was primarily over direct subsidies. Other disputes on high-technology sales encompassed public procurement, market access through trade and investment, and standards.[179]

DIRECT AND INDIRECT SUBSIDIES

Government subsidization of advanced technology has evolved quite substantially in the last twenty years. The classic industrial policy of providing direct financial support, usually in the form of grants, to national champions

---

[178] See Sylvia Ostry and Richard Nelson, *Techno-Nationalism and Techno-Globalism,* p. xviii and *passim.* The pervasiveness of the trade negotiators' agenda is such that it includes recommendations for principles and processes central to the state. In the case of China, it is argued that the absence of a transparent, enforceable, and enforced "rule of law" is incompatible with the obligations and requirements of the international economy. See Ostry, "The Post Uruguay Trading System: The Major Challenges," Industry Canada Distinguished Speakers Series, Ottawa, pp. 23–27.

[179] Sylvia Ostry and Richard Nelson, *Techno-Nationalism and Techno-Globalism,* p. 62.

or industries facing severe international competition has declined, partly due to international pressure. In countries as diverse as Germany and Japan, policymakers grew increasingly skeptical of the open-ended nature of direct support, arguing that it was expensive, ineffective, and often counter-productive.[180]

In its place, governments increasingly provide broad support for core technology activities designed to improve the competitiveness of whole sectors[181] (though in smaller economies, the distinction between sectoral support and support for individual firms is often moot). Instruments include cooperative government research contracts, government procurement (designed to provide contracts which encourage the application of cutting-edge technologies), and in some countries government-funded venture-capital mechanisms, e.g., the Japanese Key Technology Center or the U.S. Advanced Technology Program. As noted in Box B, other more traditional and widespread instruments of intervention include direct grants, loan guarantees, equity participation, targeted tax incentives, and R&D infrastructure enhancement, plus many other sectoral and regional supports.[182]

In part as a result of U.S. pressure to strengthen GATT disciplines over government subsidies, the Uruguay Round produced an agreement on Subsidies and Countervailing Measures (SCM). However, this agreement provided specific exemptions for "basic" and "applied" industrial research.[183] This agreement has been the source of some controversy, both with respect to its definitions and because many trade practitioners fear it may be a

---

[180] U.S. Technology Administration, *Key Foreign Industrial Competitors,* p. 1. This U.S. government report describes the evolution and current policies of Japan and Germany for the promotion of the competitiveness of their firms through the use of government R&D funds—policies by no means unique to Japan or Germany.

[181] Ibid, pp. 2–3.

[182] As noted in Box B, the OECD has made a sustained effort to develop a database on government policies and instruments in support of industry. The OECD data suggest that, overall, levels of support for industry are declining but the sectoral concentration is increasing, with 350 new programs in the period 1990–1993. Effective use of the subsidy notification process under the new WTO Agreement on Subsidies and Countervailing Measures should lead to an even more thorough inventory. Even "normal" government support for infrastructure, such as testing facilities, can result in competitive advantage in international competition for high-technology equipment. The possession of a French government-funded testing facility for clystron technology was described as determinant in the award of a U.S. Department of Energy contract to a French firm (also 50 percent owned by the government). See the statement by Derrel De Passe, vice president of Varian Associates, to the *Symposium on International Access to National Technology Programs, 19 January 1995.*

[183] As in any trade negotiation, objectives differed. Partly as a result of the experience with Airbus, as well as in other sectors, the United States sought to strengthen GATT disciplines over government subsidies while preserving the maximum flexibility for U.S. countervailing duty laws. Other active participants in the GATT subsidies negotiations sought the opposite. Council on Competitiveness, *Roadmap for Results,* p. 79.

source of substantial abuse.[184] Subsidies by themselves, especially developmental subsidies, are usually not first-order trade problems. Their impact is magnified and internationalized, however, when subsidies are integrated into a national policy of market protection and export targeting.

Agreement on the definition of basic and applied research is likely to prove difficult because the current Subsidies and Countervailing Measures text does not reflect current research practice. According to critics, the GATT text adopted a simple conceptualization of technological change, namely "that of representing the process as a sequence of several independent stages consisting of basic research, applied research and invention, market experimentation and commercial innovation, and lastly the diffusion of new methods and products throughout the economy...This simple linear construction ignores the wealth of empirical research at the microeconomic level which shows that technological change does not process unidirectionally through this sequence; it is a far more complicated dynamic process."[185] This distinction is relevant in terms of understanding both the innovation process itself and the inadequacy of formalistic definitions for the regulation of national technology programs.

A recent National Research Council report, though not directly concerned with GATT definitions, nonetheless underscored the inadequacy of the view inherent in the GATT text. The report notes that while some projects are clearly basic research and others clearly applied research, the discovery process often does not respect these simple categories. "Research and development are not separate, serial activities, but parallel and interdependent."[186] The flow of people, knowledge, and know-how between publicly- and privately-funded research organizations goes in both directions. Which direction in any instance depends on the industrial sector, the scientific discipline, and the type of work. Moreover, the report observes that a more severe definitional problem arises from the fact that in the United States "most federally funded research is at once both applied and basic."[187]

As an illustration, the report notes Norman Ramsey's Nobel Prize–winning work in physics, which was seminal for the development of the atomic clocks that provide timing and ranging signals key to the operation of the Global Positioning System. The development of the GPS and its associated technologies resulted from the applied research carried out by the U.S. Air Force to create a targeting and navigation system. Following the 1983

---

[184] Sylvia Ostry and Richard Nelson, *Techno-Nationalism and Techno-Globalism,* p. 60.

[185] Paul David and Dominique Foray, *STI Review,* p. 17. See also Robert White, *U.S. Technology Policy,* pp. 3–9. The author points out that in some cases, such as the airplane, the technology preceded the science. P. 3.

[186] *Allocating Federal Funds for Science and Technology,* p. 79.

[187] Ibid., p. 77.

decision to make the GPS signal available to the private sector, substantial industry investment in research and development created an array of productivity enhancing information products. Private sector product development also reduced costs and improved performance dramatically, opening the way to important new civilian applications while also contributing to increased technical capability and availability of military systems.[188]

In light of this complexity, in terms of both the research process and goals, the report reasons that a simple distinction between basic and applied research is often difficult to make and rarely decisive in defining the governmental role.[189] The lack of consensus regarding the specific approach and terminology of the Subsidies and Countervailing Measures Agreement will make its provisions for exempting R&D subsidies extremely difficult to apply. **Definitional difficulties for the current exemptions for R&D and the environment are therefore likely to become a source of international controversy. International initiatives to refine definitions and advise on disputes are unlikely to prove satisfactory.**

More fundamentally, however, the decision to provide an exemption for R&D subsidies was unwise and should be revisited. This is not to suggest that government R&D subsidies are necessarily improper. But where such subsidies distort international trade and cause injury, they should remain actionable. No reliable definition exists to separate "good" from "bad" subsidies. That being the case, the pre-Uruguay Round rule—that subsidies, of whatever type, are subject to trade action when they cause injury—provided an effective and serviceable test. Of course, governments facing the possibility of trade action may nevertheless, and quite correctly, determine to subsidize the development of new technologies.

The case for elimination of approved categories of subsidies is a strong one. First, state aid is fungible because money is fungible. Grants given for research may substitute for private funds otherwise available for that purpose. Private funds can then be freed up for applied research—or for the payment of wages or expansion of capacity, for that matter. Likewise, a $300 million grant to help comply with antipollution regulations can be used to pay wages. For the pre-1994 GATT, the question was not whether a subsidy bore a particular label, but what its effect was. Export subsidies were prohibited because they were considered to be, per se, a distortion of

---

[188] For an overview of the development of the Global Positioning System, see Supplement D. The National Research Council report *Allocating Federal Funds for Science and Technology* also cites the case of optics, one of the oldest fields in physics. Thirty or forty years ago, it was hard to see applications beyond lens design for cameras and telescopes. With the discovery of the laser and its application in fiber-optic communications, optics has turned out to be essential to modern telecommunications networks." Pp. 76–77.

[189] Ibid., p. 76.

trade flows. Other subsidies were susceptible to an offset by other countries if they were seriously prejudicial to that country or injured the latter's industry. The creation by the Uruguay Round of a safe harbor for subsidies depending on what they are called was, in fact, an unnecessary and retrograde step in the long attempt with GATT to reduce trade distortions. It was the result of a lesser concern with regulating subsidies than with the efficacy of countervailing measures.

NATIONAL SECURITY AND DUAL-USE TECHNOLOGIES

A further source of complexity regarding the role of government in supporting new technologies derives from their increasingly widespread military applications. The United States—like Japan[190]—believes its national security is based on having guaranteed, cost-effective access to the world's best technology. For many years, the U.S. Defense Department funded research and procured from specialized defense suppliers, and was often successful in sustaining both the technology and the production base at the leading edge.

The spread of what were formerly defense-only technologies has made this approach outdated.[191] Defense-only procurement cannot take advantage of the economies of scale derived from high-volume commercial production, nor can it keep pace with the rapid technological innovation of highly competitive commercial sectors. To benefit from the technological dynamism and the lower costs of commercial production, U.S. policy now is to use commercial components, technologies, and subsystems whenever possible.[192]

Under this dual-use strategy, defense initiatives will seek to create or nurture globally competitive domestic industries able to meet defense requirements by drawing on the commercial technology base.[193] (See Supplement C.) This strategy does not exclude international cooperation. On the contrary, a recent Academy report argues that tight defense budgets and the

---

[190] Richard Samuels presentation to the conference *Sources of International Friction and Cooperation in High-Technology Development and Trade, 30–31 May 1995.* See also Samuels, *Rich Nation, Strong Army.*

[191] For a cogent summary of U.S. dual-use policy, see Paul Kaminski's presentation to the conference *Sources of International Friction and Cooperation in High-Technology Development and Trade, 30–31 May 1995.* For a discussion of the challenges to current U.S. policy, see the presentation of Jacques Gansler, ibid.

[192] Flat Panel Display Task Force, *Building U.S. Capabilities in Flat Panel Displays: Final Report,* October 1994, p. I-2.

[193] Ibid.

increased use of commercially derived products suggest that Japan's technological capabilities can increasingly contribute to common security needs.[194] A key principle of the U.S. approach is that any dual-use initiative must be consistent with obligations under the WTO.[195] Nonetheless, the convergence of military need and commercial military capacity opens the door to widespread use of national security exemptions for national programs of support to industries deemed critical to national defense.

DISCRIMINATORY PUBLIC PROCUREMENT

Public procurement remains a major means of government support for national industries and a source of friction in the international system. Because a significant share of markets for high-technology products is derived from public purchases of power generation equipment, telecommunications systems and components (switches, fiber optic cable), large high-speed computers for education and national defense needs, and of course civil aircraft, competitive access to procurement markets is important for producers of high-technology equipment. The markets are large, and orders often involve significant follow-on business and can generate significant economies of scale.[196] Public procurement of high-technology products has sparked trade disputes on purchases ranging from those of Airbus aircraft by national airlines to those of telecommunications equipment by national service providers in Europe, sonar mapping equipment in the U.S., and computer equipment in Japan.[197]

Notwithstanding the attempts of the GATT to extend the reach of national treatment to public purchases, progress in opening public contracts

---

[194] *Maximizing U.S. Interests in Science and Technology Relations with Japan,* National Research Council, p. 3. To realize this cooperation, however, some adjustment in the asymmetries in capabilities and institutions for technology and industrial development will be necessary. The report notes that "these disparities have led to different preferences regarding cooperative mechanisms, with the United States in most cases preferring off-the-shelf sales of U.S. weapons to Japan, and Japan mostly preferring indigenous development." P. 4. The report identifies two reciprocal problems (which in fact capture the differing approaches of the two governments): for Japan, the report cites the "unwillingness of Japanese industry and government to cooperate technologically on reciprocal terms." For the U.S., the report identifies "the lack of consistency and coordination in U.S. government approaches" to attempts to obtain more balanced technology flows in defense technology collaboration over the last fifteen years. Pp. 4–5.

[195] Flat Panel Display Task Force, *Building U.S. Capabilities in Flat Panel Displays,* p. I-2.

[196] Bernard M. Hoekman, and Petros C. Mavroidis, *Policy Externalities and High-Tech Rivalry: Competition and Multilateral Cooperation Beyond the WTO,* OECD, Paris, October 1995, p. 18.

[197] Ibid., and Laura Tyson, *Who's Bashing Whom? passim.*

to intraregional and international competition has been slow.[198]   Impediments include the complexity and diversity of national and subnational procurement, differences in market structure, and a reluctance by officials (at the national, regional, and local levels) to transfer funds to foreign entities to the disadvantage of national producers.   Governments still continue to see public procurement as a means of supporting national champions through noncompetitive contracts.   In some cases, contracts are structured to provide the financial resources to undertake additional research in product development.   Lucrative home-market contracts can also enable firms to compete aggressively on price in third markets.

Discriminatory contracts can and do have important negative effects, not only within the trading system, but within the "protected" national economy as well.[199]   The competitiveness of downstream users of the goods and services thus procured can be constrained by their lower quality and performance, especially in relationship to the global market.   Despite these inefficiencies, given the importance of government acquisition in high-technology markets, international disputes are likely to intensify (see Supplement E).   The disputes arising from these noncontestable national markets are also likely to undermine the conditions necessary for international cooperation in the development of new technologies.

Fundamentally, a reexamination of the way the multilateral trading system addresses government procurement is now necessary.[200]   In the aftermath of the Uruguay Round agreement, government purchases are one of the few areas not covered in a thorough manner by international trade disciplines.   To a large extent, this is because the existing Government Procurement Agreement (GPA) requires its signatories to make the leap to full national treatment in government procurement, a leap that most countries remain unwilling to make.   The alternative—described in Supplement E— would adopt the GATT tariff reduction procedures as a model that could be applied to achieve steady market-access improvements in the government procurement area.   Under this approach, all WTO contracting parties would be members of the GPA, and all government-owned entities would be covered by the disciplines of the agreement.   GPA members would be allowed to give preferences to their domestic suppliers as long as they bind them-

---

[198] This discussion draws from the presentation of R. Michael Gadbaw to the conference *Sources of International Friction and Cooperation in High-Technology Development and Trade, 30–31 May 1995* and Richard E. Baldwin, *The Role of Government Procurement*, 21 November 1995 contribution to Steering Committee deliberations.

[199] Richard E. Baldwin, *The Role of Government Procurement*. Notwithstanding, governments continue to use targeted national procurement as a means of nurturing high-technology industry. See Supplement E below.

[200] For a more detailed review of potential reforms to the Government Procurement Agreement, see Supplement E.

selves not to reduce their level of openness to foreign competition. Further commitments to regular negotiations to improve the degree of procurement liberalization could be sought, and current GPA procedures could be maintained, but simplified. Retaliation, where authorized pursuant to a dispute settlement panel ruling, should be permitted through withdrawal of concessions made in any of the WTO agreements, not just the GPA, and the fundamental principle of the new agreement should be most favored nation (MFN) treatment rather than national treatment.

The fundamental issue, however, is this: increased international cooperation must imply more transparent and competitive procurement regimes. Reserving markets for national champions is ultimately incompatible with cooperative efforts to develop new technologies. This is especially true when firms benefiting from protected home markets seek access to the publicly financed technology development programs of other countries. International contestability of participating firms' home markets may increasingly become a de facto condition for cooperative activity.

PRODUCT STANDARDS

Product standards raise similar issues. Discriminatory or exclusionary standards practices are incompatible with efforts to improve international cooperation in the development of new products. At the same time, international cooperation is an excellent means to avoid conflict over differing national standards for key technologies, especially when the cooperative standards-setting is accompanied by improved transparency and mutual recognition.

In this regard, the call by the Trans-Atlantic Business Dialogue for negotiations on a full and complete mutual recognition agreement for medical devices, telecommunications terminal equipment, information technology products, and electrical equipment, as well as a common registrations dossier for new drug products, is an important new initiative that should be supported.[201]

---

[201] *Trans-Atlantic Business Dialogue: Overall Conclusions,* p. I.1. In the general remarks, the business leaders emphasized that the trans-Atlantic marketplace can flourish only against a background of political cooperation and trust. Treating the policy issues associated with high-technology competition and cooperation is an essential part of this dialogue. See, for example, the calls for greater cooperation in information technology, competition policy, government procurement, and common eligibility for R&D programs. *Passim.* For a thorough review of the standards issue, see *Standards, Conformity Assessment, and Trade into the 21st Century,* National Academy Press, Washington, D.C., 1995.

DUMPING AND ANTIDUMPING

In the imperfect competition of international trade in high-technology products, most countries have developed trade laws to counter anticompetitive practices of foreign firms.[202] Indeed, the nature of high-technology competition, particularly the possibility of achieving decisive advantage through strategic dumping, makes this practice attractive, especially to new market entrants. The issue is further complicated insofar as some instances of "forward pricing" down the learning curve can legitimately be treated as dumping and counteracted where imports cause material injury to a domestic industry in the importing country. In principle, "forward pricing" occurs when a firm prices below current costs in anticipation of generating sufficient demand to push actual costs below a price target.[203] Yet within the GATT system and the national laws implemented consistently with that system, dumping is selling below fair value, usually considered to be the home market price.[204]

Strategic dumping differs from "normal forward pricing" in that it usually involves subsidizing exports through domestic prices which are significantly higher than world market prices; it is made possible by collusive price behavior among domestic firms and restricted access to the market by competitive foreign producers.[205] Market closure may result from structural differences in industrial organization, access to capital, and corporate governance (e.g., stockholder expectations), as well as public and private restrictions on imports and investment.[206] In combination, these factors provide both solid protection and powerful incentives for firms to seek foreign market share.

Many antidumping cases brought in the 1980s defined dumping not as selling below home market price or discriminating on price in different markets, but as selling below a constructed measure of average production

---

[202] For an informed discussion of the rationale, limitations, and policy dilemmas of this controversial issue, see the section on "Subsidies and Dumping: What They Are, Why They Matter" in *Competing Economies: America, Europe and the Pacific Rim,* Office of Technology Assessment, Congress of the United States, Washington, D.C., 1991, pp. 138–154.

[203] Bernard M. Hoekman, and Petros C. Mavroidis, *Policy Externalities and High-Tech Rivalry,* p. 16. See also Michael Borrus and Jeffrey A. Hart, "Display's the Thing: The Real Stakes in the Conflict over High-Resolution Displays," *Journal of Policy Analysis and Management,* vol. 13, no. 1 (1994), pp. 50–54.

[204] Laura Tyson, *Who's Bashing Whom?* p. 267.

[205] Ostry, "Technology Issues in the International Trading System," p. 10

[206] Differences in industrial structure, expectations of capital markets, and management objectives, such as shareholder value versus market share, can result in major differences in firm behavior, in particular the willingness to tolerate sustained losses for market share. See Robert Denham and Michael Porter, *Lifting All Boats.*

cost.[207] Both methods are provided for in the GATT system and have been since its inception in 1947. Given the significant scale and learning economies that characterize high-technology industries (i.e., prices fall as output increases), there are strong incentives for producers to set prices on the basis of a future cost they hope to achieve as they increase production. National antidumping laws may reach certain pricing behavior by foreign suppliers—provided, again, that imports are causing material injury to a domestic industry—even though similar strategies remain available to domestic producers.[208]

In technology industries characterized by scale and learning economies, forward pricing strategies are indistinguishable from predatory pricing.[209] Thus, there is no basis for requiring that a dumping complaint demonstrate predatory intent, which is notoriously difficult to prove. In addition, the time required to process antidumping complaints is already the subject of legitimate concern in high-technology industries with short product life cycles.[210] Administrative delay in rapidly evolving product markets can permit foreign producers to move on to the next product cycle before a finding is made. The high cost, rapid innovation, and short product cycles characterizing these industries make possible significant damage to domestic industry in relatively short periods.[211]

---

[207] For analysis of this and related issues, see Laura Tyson, *Who's Bashing Whom?* pp. 267–272.

[208] Ibid., p. 269.

[209] Ibid., p. 268. The author reviews both the rationale of and the problems with national antidumping policies, but rejects the argument that dumping should be restricted to its original definition of selling goods for less than the home market price. She argues that it would be "both politically unrealistic and economically imprudent" to do so. She states further that "it would be imprudent to overlook the real possibility of pricing below marginal cost as a predatory business tactic, and one that is especially attractive in high-technology industries."

[210] Ibid., p. 271. Some analysts argue that, while there is room for improvement in the current application of antidumping measures, the area where reform is more urgently needed is in making antidumping measures more efficient. In high-technology competition, procedural delay can be the equivalent of denial. See Thomas Howell, "Dumping: Still a Problem in International Trade," in *Trade and Competition Policies*, Westview Press, Boulder, Colo., forthcoming. See also the section "Delay and Uncertainty in Imposing Duties," in chap. 4 in *Competing Economies*, Office of Technology Assessment, p. 143.

[211] For example, in the depressed 1985–1986 market for DRAMs, Japanese companies waged an aggressive and arguably successful price war in the United States and other markets. "This was one of the major factors behind the exit of seven out of nine American DRAM producers by 1986" and the domination of the market for the latest generation of DRAM chips by Japanese producers. Laura Tyson, *Who's Bashing Whom?* p. 101. The vertical integration of the Japanese firms and their keiretsu linkages provided access to relatively cheap and patient capital, enabling them to finance large, countercycled investments and to sustain massive losses. Ibid., p. 100. It is estimated that U.S. firms lost $2 billion during this period and Japanese firms some $4 billion. However, by the end of 1986 the Japanese firms had acquired a virtual monopoly in a key input for all the systems applications in which they were trying to become significant competitors. Ibid., p. 101.

BOX G. THE DUMPING/ANTIDUMPING POLICY DEBATE

A note is in order on the great divide existing between participants in the debate over the appropriateness of the current WTO/GATT regime of measures approved to counter dumping. The subject is a contentious one. In the United States, Europe, Canada, and Australia—all of which have strong Western legal traditions—there is in each country a domestic division between, in one camp, those in industry and government trade agencies who have experience with injury caused by dumping as defined by international agreement, and, in the other camp, academic economists and national competition authorities who vigorously object, on theoretical and philosophical grounds, to the applications of antidumping measures as an unwarranted interference in the market. Within this second camp, there are also economists and policymakers, more attuned to the realities of international competition, who nonetheless object to certain aspects of current practices, without however, rejecting the economic and political considerations on which antidumping actions are based.[212]

There is a sharp intellectual divide on the issue, partly because the natural constituencies and assumptions of the two camps differ immensely. The divide separating the two camps occurs in large part because the first proceeds on the basis of experience and a policy construct designed to address important, sometimes conflicting, national economic goals, while critics rely on economic theory, and see firms petitioning for relief as rent seekers fleeing the rigors of international competition.

The first group, that is, the industrialists and the trade policy practitioners, are much more concerned with the consequences of dumping on the current health and future prospects of the national industrial base. This includes associated industries, employment, and more broadly the technological capabilities that advanced industries bring to the national economy.[213] This group is also concerned with the political consequences of rapid declines in the fortunes of major industries.

For the most part, the trade policy group has not sought theoretical support for its position. However, it now finds a growing theoretical literature on strategic trade which calls attention to the strategic or critical nature of certain industries. These theorists argue that the targeting of specific industries under the conditions of technological externalities, increasing returns to scale, and imperfect competition can lead to

---

[212] See, for example, Laura Tyson's discussion, *Who's Bashing Whom?* pp. 267–272, noted above. The United States, of course, is by no means the only country having administrative practices which are subject to international criticism.

[213] The loss (or acquisition) of industrial capability can have important long-term consequences for the national economy. An innovative, rapidly growing industry, and its supporting infrastructure, can be a source of powerful advantage in the international industrial competition. Some analysts observe that technology develops along a path in which future opportunities quite logically grow out of the research and production undertaken today. The pace and direction of technological innovation and diffusion are therefore shaped by current mastery of production processes and market position. In this view, production costs and technology do not automatically converge among nations; they may diverge with important long-term competitive consequences. See Steve Webber and John Zysman in Borrus et al., *The Highest Stakes,* pp. 179–180 and note 33.

significant shifts in national competitive advantage.[214] From this perspective, dumping (the sale of products below marginal cost or average variable cost) may be designed to encourage competing firms to exit the market or to preempt the market by deterring other firms from entering.[215]

The political consequences of unrestrained dumping may be the most compelling consideration. Indeed, some observers suggest that a liberal international trading system based on differing economies cannot be sustained as a political matter without an "interface mechanism." These mechanisms are required in a world where

- vertical industrial organization commonly gives rise to cross-product subsidization;
- many nations base their economies on export growth strategies rather than on domestic consumption; and
- in such economies, access barriers to the national market provide above-market returns for corporations benefiting from such protection and allow them to price aggressively abroad without fear of retaliation in home markets.

In these conditions, governments adhering to a generally open trade and investment policy will not, over time, be able to garner sufficient public support to maintain open trade relations with export-oriented economies. Recognizing this "system friction," some analysts have argued that national antidumping laws "act more as a buffer between different economic systems than as a response to unfair practices."[216]

Those who oppose antidumping actions raise a number of theoretical and competitive considerations. The most theoretical critics point to increased consumer costs resulting from antidumping actions that restrict low-priced imports. Indeed, some argue that these imports constitute "a gift" to the consumers and users of the recipient countries.[217] Others argue that the pricing practices objected to in international commerce (i.e., forward pricing) are often accepted domestically. Many economists, in particular, object to the use of constructed measures of average production costs by government agencies charged with the enforcement of antidumping legislation, on the grounds that national antidumping laws can be used "to preclude foreign suppliers from

continued

---

[214] For an excellent summary of the arguments of strategic trade theorists and the related industrial policy theories, see Jeffrey A. Hart and Aseem Prakash, "Implications of Strategic Trade for the World Economic Order," paper prepared for the Annual Meeting of the International Studies Association, San Diego, Calif. 16–20 April 1996. Historically, dumping became a problem in international trade only around 1880, when the manufacturing enterprises of two newly industrializing powers, Imperial Germany and the United States, began mounting a challenge to British commercial hegemony using dumping as a systematic export tactic. See Jacob Viner, *Dumping: A Problem in International Trade,* University of Chicago Press, Chicago, 1923. The author helped draft the U.S. Antidumping Act of 1921.

[215] See section on Dumping and Antidumping in text. See also Laura Tyson, *Who's Bashing Whom?* p. 268.

[216] See John H. Jackson, The World Trading System. MIT Press, Cambridge, Mass., 1989, pp. 220–221. See also Laura Tyson, *Who's Bashing Whom?* p. 270.

[217] Gifts are rare in international commerce, even more so in relations among nations. Consumers have to have income to profit from low-priced goods. "In the long run, if you lose jobs and income as a result of predatory trade practices, consumer welfare will be lower." See

using competitive tactics that are permissible to domestic suppliers."[218] Lastly, many analysts object to the seemingly arbitrary judgments that are required of government agencies forced to reach decisions on the basis of complex and contradictory information in relatively short timeframes. In addition, while many companies are agnostic about the principles of antidumping actions, they are quick to recognize and object to government actions which raise the costs of key inputs. In these cases, aid to one industry, or segments of an industry, can harm the international competitiveness of a user industry, and even result in the transfer of domestic production to offshore sites.[219]

The more theoretical critics of antidumping policies are unconcerned that they are joined in their criticism of antidumping policies by nations which have substantially less open economies (in terms of trade, investment, and technology flows) and which deploy a wide variety of non–GATT-sanctioned means to regulate import competition. In principle, these critics would argue—with logic unassailable at a theoretical level—that multilateral solutions should be adopted to remove these barriers. This approach has merit; indeed it is advocated elsewhere in this analysis. However, such solutions are necessarily long-term and by no means certain of success. Consequently, proposals for long-term solutions do not adequately address the immediate competitive challenges faced by both the affected industry and national policymakers.

Given the widely differing perspectives present in the contention over the dumping/antidumping issue, the debate is not likely to be resolved, nor are these sharply divergent views likely to be bridged. Conduct and practices deemed economically irrational by critics are seen as essential for the protection of strategic industries by those responsible for the future economic capabilities of their country and the health of the world trading system.

---

the presentation "Consequences for the International Economic System" by Lawrence Chimerine in C. Wessner (ed.), *Sources of International Friction and Cooperation in High-Technology Development and Trade.*

[218] Laura Tyson, *Who's Bashing Whom?* p. 269. For a generally critical assessment of American antidumping laws, see Richard Boltuck and Robert Litan (eds.), *Down in the Dumps: Administration of the Unfair Trade Laws,* The Brookings Institution, Washington, D.C., 1991. Many of the criticisms of U.S. antidumping measures in this volume are by lawyers who represent respondents, i.e., foreign companies charged with dumping. In the same volume, Michael Coursey, a former Department of Commerce official, argues that "most of this criticism seems to be much ado about nothing." He points out that no specific examples of trade law injustice, i.e., wrongful convictions of dumping, are cited, "because, for example, weighted average foreign prices were used..." Ibid., p. 239. For a critical assessment of European antidumping policies, see Patrick A. Messerlin, "Reforming the Rules of Antidumping Policies," paper presented at the conference *Toward A New Global Framework for High-Technology Competition, 30–31 August 1995.* Antidumping policies can play a crucial role in defending high-technology industries but, as Messerlin points out, most antidumping cases deal with products which would not be considered high-technology industries; many cases involve products such as steel plates, barbed wire, bicycle chains, PVC, textiles, Portland cement, and soda ash. P. 2.

[219] Council on Competitiveness, *Roadmap for Results,* p. 26. See also Hart, "Anti-Dumping Petition of the Advanced Display Manufacturers of America."

Moreover, the higher returns which may accrue to national firms benefiting from these practices can provide the resources to fund additional research, more rapid product development, expanded marketing, and overseas acquisitions of competitors. Even when practiced for relatively short periods, these strategies provide substantial competitive advantage in high-technology markets. For the recipients of dumped products, the revenue losses from reduced exports and domestic market share are compounded by the loss of the dynamic efficiency gains (i.e., learning by doing) that characterize high-technology industries. The cumulative effect of these practices can permanently alter the terms of international competition by forcing competing firms to exit a product market or by deterring new entrants.[220]

In these circumstances, the need for prompt and effective antidumping policy at the national level is heightened. This may be a second-best policy solution, but it is likely to prove essential for countries with relatively open markets in high-technology goods.[221] From the international perspective, unilateral national action could be usefully supplemented by improved consensus and standards on competition policy and its enforcement.

MARKET ACCESS: COMPULSORY TECHNOLOGY TRANSFERS AND AEROSPACE COMPETITION

In the intense competition for sales of high-technology products in telecommunications, power plants and aircraft, companies from both Europe and the United States find themselves engaged in competition for market share. Product quality, such as performance, reliability, and life-cycle cost, are major determinants of commercial success, but not the only determinants. As noted above, discriminatory public procurement practices can

---

[220] Paul Milgrom and J. Roberts, "Limit Pricing Under Incomplete Information," *Econometrica,* vol. 50, no. 2, March 1982, pp. 443–459. See also Paul Milgrom, *Predatory Pricing,* 1987, cited in Laura Tyson, *Who's Bashing Whom?* p. 268. Even the protection offered by antidumping laws may prove inadequate to protect an industry, first because the imposition of duties is slow and uncertain and second, even when promptly applied, these duties may often be inadequate to neutralize the advantage gained by foreign competitors. *Competing Economies,* p. 143.

[221] Ending dumping in U.S. and third-country markets is considered one of the achievements of the Japan-United States Semiconductor Agreement. See Ministry of Foreign Affairs, Japan, *Salient Points and Data Related to the Japan-U.S. Semiconductor Arrangement,* Tokyo, February 1996, p. 1. For a fuller discussion of this point and the impact of the agreement, see Supplement A.

play a decisive role in the purchase of capital-intensive goods such as turbines, aircraft, and defense-related equipment. Many countries seek concessions, especially for public purchases of aerospace and defense equipment, often called offsets.[222] A recent U.S. government report noted that, "in defense trade, offsets include mandatory co-production, license production, sub-contractor production, technology transfer, counter trade and foreign investment."[223] The report concludes that, "as the world's preeminent supplier of weapons...civil aircraft and high cost-high technology hardware, U.S. corporations are highly vulnerable to offset demands" and notes that this practice poses a problem for future economic security.[224] Leading high-technology corporations from Europe face similar demands—and risks.

The aircraft industry, with its slow production cycle, large unit costs, and long-term maintenance and supply contracts, is a prime example of both this fierce high-technology competition and the growing use of offsets as a means of compulsory technology transfer. The global competition for aerospace exports is increasingly exploited, particularly by the steadily more capable economies of East Asia. Countries such as Japan, China, and South Korea demand production offsets to reduce imports and augment their technological capabilities in the aerospace sector. Agreeing to offsets in order to gain access to Asian and other markets has become a prerequisite to sales of commercial aircraft. Offsets frequently occur in countries such as China that retain state control of the economy but are frequent in advanced economies as well where this practice is often facilitated by the

---

[222] The term "offset" is frequently associated with military sales where the costs of weapons systems, e.g., airplanes, to a government is reduced, i.e., offset, by a certain amount through work performed in the home country with employment and other spinoff benefits. Both the EU countries and Japan have traditionally required offsets for military equipment; the practice now extends to high-technology commercial contracts.

[223] Bureau of Export Administration, *Offsets in Defense Trade*, U.S. Department of Commerce, Washington, D.C., May 1996. The report notes that offsets may be direct or indirect. The former refer to coproduction or subcontracting directly related to the exported system; indirect offsets refer to compensation unrelated to the export, such as foreign investment.

[224] Statement by William A. Reinsch, Commerce Under Secretary for Export Administration in BXA-96-8, 17 May 1996. The BXA report notes that "the newer offset customers, especially in the Middle East, are seeking to diversify their economies rather than build or maintain a defense industry. Pacific Rim countries such as Singapore, Taiwan, and South Korea are seeking offset deals that include increased technology transfer, particularly in aircraft design, to become self-sufficient in military production and to overcome industrial weaknesses that are hindering their efforts to compete in the world aerospace market with U.S. and European manufacturers. Japan's policy of co-producing defense items has a similar objective." Bureau of Export Administration, *Offsets in Defense Trade*, p. 6.

important government role in state-owned airlines and, of course, military aviation programs.[225]

The aerospace industry thus provides a clear, even stark, example of the stakes of international competition for high-technology industries and the high-wage jobs they bring. A recent U.S. study notes that "commercial aerospace has become a key target industry for many advanced as well as industrializing nations,"[226] who have adopted explicit industrial policies to advance this goal. Countries as diverse as Germany, France, Japan, and China believe the aerospace industry has the broad range of spillover effects typical of high-technology industries. (See Box F.) Accordingly, they seek to learn the organizational and design skills, build the supporting infrastructure, and acquire the high-skilled jobs and high-value exports that characterize this industry.[227]

The extent of governmental involvement in this sector, combined with the importance of downstream employment effects and the limitations on the demand for aircraft, suggests that aerospace is a sector that will continue to generate substantial friction. In this regard, international competition for the benefits of a national aerospace industry, especially high-wage employment, captures the conflicting trends towards globalization and national competition that characterize some high-technology industries.

---

[225] A recently announced Japanese effort to develop a new antisubmarine warfare patrol aircraft in the 1996–2000 Midterm Defense Program illustrates the close government involvement in aerospace development and may be a source of future U.S. friction with Japan. The Japanese press report notes that "there is a strong possibility that the U.S. defense industry, which is suffering from a sharp decrease in U.S. forces procurement, will try to intrude into the program by influencing the U.S. government and Congress." The Japanese editorial expects "a debate greater than the one at the time of the FSX." An intriguing feature of the proposed ASW aircraft is that "the aircraft's outer form and size are likely to be similar to those of the twin-engine passenger plane Boeing 737, which are used by local airlines." Tokyo AERA article by AERA Board Member Shunji Taoka, "The Whole Picture of the Domestic Patrol Aircraft Plan," 12 February 1996, pp. 16–18.

[226] Randy Barber and Robert E. Scott, *Jobs on the Wing: Trading Away the Future of the U.S. Aerospace Industry,* Economic Policy Institute, Washington, D.C., 1995, p. 6. Sylvia Ostry, "Technology Issues in the International Trading System," also notes that improving the balance of technology flows is a high priority for a growing number of countries, adding that these national "efforts to induce [technology] inflows and reduce outflows are bound to lead to disputes and also will reduce global welfare" (p. 11). This is true. It is also true that there will be winners and losers, i.e., some nations will gain technological capability at the expense of others. Laura Tyson, *Who's Bashing Whom?* p. 40.

[227] Barber and Scott, *Jobs on the Wing,* p. 6. Similar motivations underpin the proliferating national programs to build indigenous microelectronics industries.

BOX H. THE STAKES IN AEROSPACE COMPETITION

Aerospace has been identified as a prime target of national and regional industrial policies around the world. A number of these competitors have the goal of developing a full-service commercial aerospace industry. China is assembling entire Western-designed jetliners. German aerospace efforts have been consolidated under the umbrella of Daimler-Benz (although the cost and commercial challenges of this competition are highlighted by the bankruptcy of Folker). Japan is mounting a systematic effort to become a first-tier aerospace manufacturing power, an approach underscored in a recent National Research Council review of U.S.-Japan technology linkages, appropriately entitled **High-Stakes Aviation.**[228] Even only recently industrializing countries such as Indonesia have joined over thirty other participants in the global contest for a share of aerospace.[229]

In Europe, the participants in the Airbus consortium have successfully reestablished a European role in commercial aircraft production. To compensate for the advantages they perceived as accruing to American aircraft producers through years of direct and indirect (primarily military) support, the participating governments provided direct financial and commercial support to their national participants in the consortium. This direct support included government contracts for development, loans and loan guarantees on favorable terms, equity infusions, tax breaks, debt forgiveness, and even guarantees against losses caused by changes in exchange rates.[230] Overall, the effort to build Airbus involved years of sustained expenditure, with some estimates of the **net** European investment at about $26 billion.[231] In addition, European governments were able to influence procurement decisions of national airlines, through both outright government ownership and regulatory powers.[232] This preferential procurement provided the crucial, first initial launch orders necessary to begin aircraft programs.[233]

This sustained, integrated, regional effort has been a success.[234] The Airbus consortium is now the world's second-largest commercial aircraft manufacturer and supports

---

[228] *High-Stakes Aviation: U.S.-Japan Technology Linkages in Transport Aircraft,* National Academy Press, Washington, D.C., 1994.

[229] Ibid., p. 87.

[230] For a discussion of the complexity of the Airbus accounts, see *Competing Economies,* Office of Technology Assessment, p. 353. Airbus itself receives no financial support; "the disbursement and repayment of launch aid and other supports is handled among the members and their governments." Ibid.

[231] Gellman Research Associates, "An Economic and Financial Review of Airbus Industrie," 1990, cited in Barber and Scott, *Jobs on the Wing,* p. 50.

[232] *Competing Economies,* p. 355.

[233] Ibid., pp. 354–357. European analysts argue that "cost-plus" U.S. military contracts have provided high rates of return for U.S. aircraft manufacturers on military programs, thereby enabling them to initiate commercial aircraft programs whose risk-adjusted prospective rates of return are below "ordinary" hurdle-rates. In addition, the contract price of military aircraft programs generally includes funding for manufacturing facilities; in some cases, military programs helped defray initial design costs. See H. Landis Gabel and Damien Neven, "In Defense of Airbus Industrie" in Olivier Cadot et al., *European Casebook,* pp. 178–181. During the post-war period, the U.S. commercial aircraft industry also benefited from substantial direct support through NASA and NACA and from indirect financial support through military research. See David C. Mowery and Nathan Rosenberg, *Technology and the Pursuit of Economic Growth,* Cambridge University Press, New York, 1989, p. 188.

[234] Not all analysts consider Airbus a success. Indeed, by some measures, public costs may outweigh gains. See, for example, Damiel Neven and Paul Seabright, "European Industrial Policy: The Airbus Case," *Economic Policy,* no. 21, October 1995. The authors find that,

over 100,000 jobs, with major spillover effects for European economies, especially the development of the European aerospace supply base. Given the success of the Airbus program and the broader benefits to the economies of consortium members, it is not unreasonable to expect that the prospect of similar success will eventually attract new, state-supported market entrants.

Aerospace has become a key target industry for both the advanced and industrializing nations. They seek to acquire the industrial infrastructure and skills associated with "aircraft electronics, advanced metal fabrication, composite materials, turbine engines, and other (aircraft) components."[235] They recognize that aerospace has broad spillover effects and that "the organizational skills and discipline to design, fabricate, assemble, market and service commercial aircraft"[236] are of great national benefit, not only as high-value exports but also as a platform for other high-technology industries.

The competition for this sector shows no signs of diminishing in its intensity or in the level of direct government support. A recent report by the National Research Council concluded that heightened international competition and dramatic reductions in the defense budget mean the U.S. industry "will be severely challenged over the next decade just to hold its current position in global aircraft manufacturing."[237] Other recent reports call for a series of measures to maintain U.S. technological leadership, revitalize U.S. manufacturing, ensure "a level playing field for international competition," and develop "a shared U.S. vision" of the industry.[238]

The belief that meeting this challenge will require government action is widespread. As one analyst observed, "rather than hope that foreign intervention in the industry will cease, the United States will have to respond with offsetting interventions at home. Rather than assume that dependence on foreign suppliers does not matter, it will have to recognize the dangers of excessive dependence on such suppliers for critical component technologies, such as advanced avionics."[239]

This observation was prescient. The Japanese government strategy of funding development costs, as leverage to obtain work-shares and access to leading-edge technologies, has had substantial success.[240] As suppliers of critical subcomponents and assemblies, the major integrators have become sufficiently dependent on Japanese firms

continued

---

given the prior presence of McDonnell-Douglas, entry by Airbus reduced airline prices only modestly. Hence, they find the consumer surplus argument for government subsidy to Airbus is weak, given the competition already provided by MDD. This analysis undercuts the justification for government support on consumer welfare grounds. However, it does not address the strengthening of the European aerospace industry, including employment, through the shift in rents from U.S. producers to European producers. See *Competing Economies*, p. 352. For a consumer-oriented analysis critical of the Airbus program, see Richard Baldwin and Paul Krugman, "Industrial Policy and International Competition in Wide-Bodied Jet Aircraft" in Richard E. Baldwin (ed.), *Trade Policy Issues and Empirical Analysis*, University of Chicago Press, Chicago, 1988.

[235] Barber and Scott, *Jobs on the Wing*, p. 6.

[236] Ibid.

[237] *High-Stakes Aviation*, p. 75.

[238] Ibid., p. 5.

[239] Laura Tyson, *Who's Bashing Whom?* p. 211.

[240] See *Competing Economies*, pp. 349–352. The report cites a MITI report which states the need for the government to subsidize aircraft development projects and develop aircraft engineering "on the initiative and the assistance of the government as it involves highly sophisticated and complex technology." "The Vision of MITI Policies in the 1980s," cited in David C. Mowery and Nathan Rosenberg, *The Japanese Commercial Aircraft Industry Since*

that Boeing claims it can no longer build an aircraft without the participation of Japanese suppliers.[241] Japan has made a similar sustained effort to develop its capability in military aircraft.

The FSX is now entering service as the F2, representing the successful, if costly, completion of this controversial cooperative program under which some 50,000 different technologies were reportedly transferred to Japanese manufacturers by the United States. The joint development program also called for the transfer of leading-edge Japanese technologies to U.S. manufacturers.[242] Some reports, noting the aircraft cost almost twice what the Japanese government expected to spend, have described the plane as a "sleek, supersonic monument to the inordinate cost of Japan's industrial policy—the strategy of entering new industries by subsidizing businesses to develop them."[243] Others see this co-development project both as a source of momentum for Japan to develop its defense and commercial aerospace industry, and as a source of pressure to ease its ban on weapons exports in order to gain the economies of scale necessary to make systems like the FSX economically viable.[244]

Continued U.S. job losses and deterioration in the import-export ratio are causing considerable concern among U.S. aerospace workers. (By some estimates, the combined effects of reduced defense spending and the steepest recession in airline history have resulted in the loss of over 500,000 aerospace jobs in the last five years.[245]) In this environment, pressure for U.S. policy initiatives is mounting, backed by strong labor support, with calls for U.S. measures to counteract foreign industrial policies, such as financial inducements, offset agreements, and subsidies.[246] There are also calls to coordinate U.S. aerospace policy; promote aerospace production and employment— not least by the elimination of policies which work against U.S. job creation (e.g., antitrust barriers permitting U.S. companies to team with foreign companies, but not with other American companies); and negotiate new international trade agreements, including a new civil aircraft code in the WTO which would forbid offsets. These nascent U.S. efforts to adjust to the new competitive environment, especially the drop in defense spending, in conjunction with the impact of the industrial policies of other nations, suggest that this area may well be a source of future international friction.

1945: Government Policy, Technical Development, and Industrial Structure, The International Strategic Institute, Stanford, Calif., 1985.

[241] Barber and Scott, *Jobs on the Wing*, p. 59.

[242] See, for example, Steve Glain, "Concern Over 'Menace' Dissipates as Japan, U.S. Unveil Fighter Jet," *Wall Street Journal*, 22 March 1996, p. A1. Critics who note that the Japanese government is paying five times as much for an F-16 as for a plane U.S. experts consider not much more capable miss the point, which is not whether the FSX is successful in terms of the performance of the plane itself—though Japanese Air Force officials argue it is superior in important respects. The major Japanese goal in undertaking the project is the mastery of the technologies and systems, e.g., avionics and fire control, and the manufacturing skills necessary to produce an advanced fighter aircraft. Ibid. For a detailed account of the motivations and negotiations surrounding the FSX controversy, see Jeff Shear, *The Keys to the Kingdom,* Doubleday, New York, 1994.

[243] Steve Glain, *Wall Street Journal.*

[244] Ibid.

[245] Barber and Scott, *Jobs on the Wing,* p. 1. The report summarizes research which forecasts a loss of $129 billion in sales to foreign producers between 1994 and 2013, with substantial increases in imports during the same period. The report estimates that, as a result of this increased foreign competition and imports, up to 250,000 jobs in aerospace and related industries are at risk over the next four years. Ibid., p. 2.

[246] Ibid., p. 74.

INTELLECTUAL PROPERTY PROTECTION

Most high-technology industries are distinguished by the importance of intellectual property rights for research and discovery of new processes and products. The deployment and exploitation of intellectual property depend on the ability to maintain ownership and control over their inventions. Most industrial countries have developed a comprehensive and sophisticated system of protecting intellectual property rights (IPR). However, there remain significant divergences among the United States, the European Union, and Japan. This divergence is especially clear in patents where some believe that the U.S. system favors the creation of intellectual property over its diffusion, whereas other systems tend toward greater diffusion of the property thus created. International calls for harmonization of IPR regimes are likely to increase, although prospects for progress remain uncertain.

In addition to efforts at harmonization, intellectual property protection faces two other international challenges. First, there is a need to promote effective worldwide protection of intellectual property rights, particularly in rapidly industrializing countries. Many of these countries have laws designed to protect intellectual property; however, for various reasons, effective enforcement has remained elusive. As a result, high-technology industries incur substantial losses in these markets.

For example, U.S. industries estimate losses of $2.2 billion in 1995, more than double the $1 billion lost in 1994 as a result of intellectual property theft in the People's Republic of China alone.[247] A single CD-ROM illegally produced in China can hold $10,000 of U.S. software and sell for less than $10 in Hong Kong. Countries such as Brazil, India, and Korea reportedly have similar intellectual property enforcement problems. Such high–value-added industries as motion pictures and pharmaceuticals are impacted by inadequate enforcement, particularly as the share of their revenues generated by sales in international markets increases.

While each country has its own enforcement obstacles, there are common themes. In many instances, particularly in larger countries such as China, it may be difficult to secure cooperation from regional and local officials

---

[247] See Statement of Ambassador Michael Kantor before the Senate Subcommittee on East Asian and Pacific Affairs, pp. 4–5. See also Scott Williams, "Anti-Piracy Groups Target China—Technology Companies Want Federal Action," *Seattle Times,* 18 February 1994, p. E1. Intellectual property protection remains at the forefront of the international trade agenda. For example, the U.S. government considers the inadequate protection of intellectual property rights to be one of the most important issues on the bilateral trade agenda with China. Despite progress under the terms of a 1992 agreement creating a sound legal regime for IPR, its enforcement has been inadequate in areas ranging from computer software, motion pictures, and sound recordings to pharmaceuticals and agrichemicals. See Kantor testimony, 7 March 1996, *passim.*

outside the central government. Furthermore, in many countries which pursue a policy of promoting diffusion of technology, lax enforcement serves as an economic incentive that encourages intellectual property theft on a grand, even industrial scale. Some software producers have dubbed certain countries "one-copy" countries, meaning that once a single copy of a program gets into the country, practically all further copies in the market are illegal copies.[248] It is important to note that the problem of illegal copies is not confined to the Chinese domestic market. Mainland pirate producers, transshipping through Hong Kong, are exporting illegal copies of CD-ROMs and CDs to markets in Southeast Asia, Latin America, North America, and Russia and the CIS states. Moreover, the problem seems unlikely to diminish as China continues to import CD production lines.[249]

The second challenge is rapid technological change in high-technology areas such as global information systems and biotechnology. These sectors raise important new issues which may require significant adjustment to existing IPR rules. For example, the global explosion in Internet usage—as well as its potential as a channel for the export of ideas and intellectual property—raises serious government policy considerations. Internet-related legal issues such as the scope of copyright protection afforded transmitted digital information, as well as the ability of foreign governments to censor information made available on the Internet, are just beginning to emerge on a global basis. At this time, technology appears to have outpaced the development of existing intellectual property regimes. Even fundamental questions, such as which country's copyright laws should apply—the country in which the "client" computer (the purchaser of the information) is located or the one in which the "server" computer (the provider of the information) is located—have not been resolved.[250]

Similar dramatic technological changes exist in the biotechnology field, particularly in the area of genetics. Intellectual property law is being called

---

[248] "U.S./China Make Progress in Talks; Many Industries Urge U.S. Action." *International Trade Reporter,* vol. 12, no. 4, 25 January 1995, p. 168.

[249] The U.S. government has identified some thirty-four CD factories in South and Central China, producing some 54 million CDs and LDs in 1995 for a domestic market that can absorb only two to five million. Continued imports of CD production lines suggests that China may soon have the production capability that is close to 200 million CDs. Kantor testimony, 7 March 1996, p. 4.

[250] Mark Turner, "Labyrinth of Laws Could Lead to a Net Loss," *The Independent,* 15 January 1996, p. 11. Recently, a controversial German court decision effectively forced Compuserve, a U.S. Internet access and information service provider, to shut down certain "pornographic" files for all global users because, under German law, such files were illegal. While Compuserve is reportedly preparing a legal challenge, there appears to be no body of law or precedents which will be helpful in resolving the dispute.

upon to extend traditional patent protection to genes and perhaps laboratory-created organisms. Hanging in the balance of such disputes are rights to diagnostic tests and treatments for diseases (such as breast cancer), which have a tremendous monetary value. Such advances will require new norms, best achieved through international accords.[251]

At the national level, patent holders face substantial hurdles to useful enforcement because of the slow and cumbersome movement of patent litigation. The length and cost of enforcement procedures can, in effect, deny protection to patent holders who lack the resources to defend patent rights. Practices such as "blanket filing" of patents can pose nearly insurmountable problems for small undercapitalized companies. No matter how innovative, small companies cannot afford the protracted legal battles necessary to defend their intellectual property and can therefore be forced to license it. The result is that "a new competitor is created without having the research and development costs of the original innovator."[252] This can mean that the smaller company finds itself at a substantial disadvantage in selling a technology it invented and brought to market.[253] In these circumstances, that is, competition between a small innovative company and a large resource-rich multinational with superior capability in litigation, marketing, and distribution, the financial pressures for accommodation through licensing can become compelling.

Even for major U.S. companies, ensuring effective intellectual property protection and meeting the cost of vigilant enforcement can be daunting tasks.[254] Indeed, in the late 1980s, even companies as robust as Intel argued

---

[251] *Trans-Atlantic Business Dialogue: Overall Conclusions,* p. II.4. See also Ostry, "Technology Issues in the International Trading System," pp. 11–12.

[252] A 24 June 1988 letter submitted to the Senate by Intel in support of Fusion Systems Inc. technologies testimony detailing Mitsubishi's efforts to lay claim to its microwave lamp technology, cited in an article by Skip Kaltenheuser in *New Technology Week,* 5 February 1996, p. 7. The Fusion Systems congressional testimony argued that large Japanese companies are able to manipulate their patent system as a result of the "absence of a strict duty to disclose prior art—an absence that precludes any need to demonstrate why the claim in an application represents a true inventive step." Fusion Systems also argued that, because "there is no effectively enforced patent fraud statute," there is "no danger of penalty for applicants...who knowingly copy...technology" and file patents on it. Ibid., p. 3. The *1996 U.S. Government National Trade Estimate* report finds that foreign companies continue to face a series of obstacles in trying to obtain effective patent rights in Japan, including narrow claims and patent interpretation, patent-flooding, and a difficult and very slow judicial appeals process, among others. *1996 National Trade Estimate Report,* USTR, p. 187.

[253] Skip Kaltenheuser in *New Technology Week,* op. cit.

[254] Robert M. White, *U.S. Technology Policy: the Federal Government's Role,* p. 19. The author cites the defense by Corning Glass against patent infringement by Sumitomo, first in exports to Canada, then directly to the United States, and finally in a U.S. production facility. Corning's defense of its sixteen-year, several hundred million dollar investment illustrates the importance of effective intellectual property protection as well as the need for the resources to execute an effective defense.

that patent flooding "makes the standard for competitiveness in Japan a firm's size and ability to fund litigation rather than innovation, productivity and marketing."[255] Even if, as some argue, these differences in intellectual property regimes result from different historical development and current national needs, differences over intellectual property norms and enforcement are likely to prove a growing source of international friction. Efforts at harmonization of intellectual property protection must strike a balance between legitimate desires for both protection and diffusion. Without adequate protection, however, there will be less innovation to diffuse.

In addition to work on harmonization, there is a pressing need for sustained international commitment to the protection of existing intellectual property rights. At the same time, a cooperative effort should be undertaken to adjust existing intellectual property rules to the needs of new fields such as global information systems and biotechnology. Multilateral efforts to address these issues constructively should be encouraged.

INVESTMENT INCENTIVES

The political attractiveness of investment in high-technology industries and the associated high-wage, high–value-added jobs leads both nation-states and regional subunits into a beggar-thy-neighbor competition, often through generous investment incentives. In the U.S., incentive programs can be grouped into three broad categories: tax incentives, financial aid, and employment assistance.[256] These incentives include tax subsidies, infrastructure development, and other incentives, e.g., grants in kind, worker training, loans, and extensive tax holidays. Many American states have adopted industrial policies focused on this locational competition. As a result, "the financial incentives offered by competing States have grown dramatically since the mid-1970s."[257] The rapid growth in the locational competition among states and the heavy subsidies this competition engenders have sparked calls for federal action.

---

[255] Skip Kaltenheuser in *New Technology Week,* p. 7.

[256] See Chris Farrell in Fettig (ed.), *The Economic War Among the States,* Federal Reserve Bank of Minneapolis, May 1996, p.1.

[257] *Multinationals and the National Interest,* Office of Technology Assessment, pp. 66–67. The amounts involved in these incentive programs continue to grow. In 1992, the state of Ohio paid $16 million in direct incentives to Honda. (Ibid.) More recently, for a new vehicle production facility, Daimler-Benz reportedly received a total package estimated to be between $250 and $300 million, i.e., nearly as large as the much-debated Advanced Technology Program of the Department of Commerce. For an analysis of the policy issues concerning the competition among the industrial policies of state governments in the United States, see David Fettig (ed.), *The Economic War Among the States.*

Those favoring federal action are concerned that the states are engaged in a competition which, in the aggregate, is economically destructive, particularly in an era of limited public resources. Federal regulation in the United States would emulate the role of the European Union, which has sought to impose discipline on the investment incentives offered by European member states. These incentives are linked to regional economic conditions and are subject to review under EU jurisdiction over competition policy. Others argue that as cities, states, and regions increasingly become the locus of competitive advantage, it makes little sense for national authorities to limit their flexibility in attracting investment, an argument that can be applied to European nations as well as the American states.[258]

With respect to the locational competition among American states, some observers have identified an emerging consensus on several aspects of the competition among states (which are not unrelated to international competition). These potential areas of agreement would include (1) the disclosure of the full cost of incentive programs; (2) mechanisms for tracking the performance of public-supported investments; (3) the strategic targeting of investment incentives toward depressed regions; (4) a reduction in public expenditure on multinational corporations and sports franchises; and (5) greater investment in upgrading worker skills to meet the business needs of local entrepreneurs.[259] Interestingly, one final area of agreement is noted, namely that "no state can stop using development incentives in a world of fierce domestic and international competition. To do so unilaterally would be politically and economically suicidal."[260] This belief is, of course, not confined to the economic competition among American states.

Better understanding of the scope of these incentives, and of their aggregate costs and benefits including their impact on decisionmaking at the level of the firm, would be very useful to national and regional policymakers alike. Within national jurisdictions it may be possible to establish norms—with legislative underpinnings if required—to limit excessive public-private transfers in the competition for new investment. At the international level this subject should be explored as a potential element for inclusion in the negotiations for the multilateral agreement on investment. (See the discussion of national investment regimes below.)

---

[258] See Graham S. Toft in David Fettig (ed.), *The Economic War Among the States*, p. 10.
[259] Chris Farrell in David Fettig (ed.), *The Economic War Among the States*, p. 4.
[260] Ibid.

Different National Investment Regimes and Their Consequences

NATIONAL INVESTMENT REGIMES DIFFER

Inward foreign direct investment, especially in high-technology industries, can bring substantial benefits to the host country. For example, in the United States the rapid expansion of foreign direct investment has helped compensate for the low rate of domestic U.S. savings through both large numbers of acquisitions and the construction of greenfield distribution and manufacturing facilities. Investment by technologically advanced companies can also bring new technologies, processes, products, managerial skills, and organizational techniques to joint venture partners, suppliers, and ultimately competitors. New companies also create substantial employment, often in sectors with higher-than-average wages and value added. The skills and training imparted by such employment have further positive spinoff effects throughout the national economy. In addition, improved products and services can mean greater benefits for host-country consumers.

Reflecting these views, countries such as the United States and the United Kingdom have maintained open investment regimes. With some exceptions, albeit in major areas such as telecommunications, civil aviation, and broadcasting, the U.S. investment regime is among the most open on the planet, with the principle of national treatment generally applied.[261] For many policymakers in the United States, the benefits of foreign direct investment are an article of faith and, as noted above, there is considerable justification for this view.[262] It is significant, especially with respect to high-technology competition, that so few of the advanced industrial countries and the rapidly industrializing countries seem to have the same confidence in this policy approach.

In fact, many influential participants in the world trading system have quite different investment regimes, particularly with respect to the most

---

[261] *Multinationals and the National Interest,* p. 48. In 1991, the Bush administration affirmed that "the United States has long recognized that unhindered international investment is beneficial to all nations, that it is a positive-sum game." *Economic Report of the President,* U.S. Government Printing Office, Washington, D.C., February 1991, p. 262.

[262] Ibid. For a comprehensive review of the issues associated with foreign direct investment, see Edward Graham and Paul Krugman, *Foreign Direct Investment in the United States,* Institute for International Economics, Washington, D.C., 1996. See also David Bailey, George Harte, and Roger Sugden, "U.S. Policy Debate Towards Inward Investment," *Journal of World Trade,* vol. 26, August 1992, pp. 65–93.

technologically advanced industries.[263] While national investment regimes vary considerably by country and sector, none is as receptive to foreign direct investment in high-technology sectors as the United Kingdom and the United States. Despite attempts by the OECD to win acceptance for the principle of national treatment, countries as diverse as France, Italy, and Germany, and to a greater extent Japan, either formally regulate foreign direct investment or permit informal barriers which impede it.[264]

Among the advanced industrialized countries, Japan is by far the most restrictive. For many years, structural and cultural obstacles joined Japanese government policy in making foreign investment, especially acquisitions, exceedingly difficult.[265] Unfortunately, the situation remains largely

---

[263] Many countries, such as Japan, France, Taiwan, Canada, Korea, and Australia, require government notification and/or the screening of high-value investments. While formal prohibitions on the acquisition of domestic firms by foreigners have declined, many governments retain the power to review—and effectively block—foreign acquisitions, a capability buttressed by exclusionary private business practices. See Defense Science Board, Industrial Base Committee, *Report of the Defense Science Board Task Force: Foreign Ownership and Control of U.S. Industry,* Washington, D.C., June 1990, pp. 29–30. The United States also limits foreign investment in sectors such as telecommunications and nuclear energy. See James K. Jackson, *Foreign Direct Investment in the United States,* U.S. Library of Congress, Congressional Research Service, CRS Report IB93011, Washington, D.C., 1993.

[264] The European countries are more open to greenfield investments; acquisitions are more difficult. Reich emphasizes the competitive consequences of these differences in investment opportunities, noting that "Anglo-American firms have often encountered a different pattern of regulation when investing abroad. They have often been forced by host governments to invest in fully-integrated production facilities, exchange market access for patents, or have often been denied any investment access at all. Recent evidence suggests, for example, that a series of 'structural barriers' continue to deny U.S. firms the kind of reciprocal access to some foreign markets that their rivals enjoy in the United States." See Simon Reich, "Asymmetries in National Patterns of Foreign Direct Investment: Consequences for Trade and Technology Development," p. 31, in C. Wessner (ed.), *Sources of International Friction and Cooperation in High-Technology Development, Competition, and Trade.*

[265] This is not to say that foreign investment in Japan is impossible. However, successful foreign investments remain the exception. These constraints are outlined in detail in the second annual working report of the U.S.-Japan Working Group on the Structural Impediments Initiative. Companies such as IBM, Texas Instruments, and Motorola have penetrated the Japanese market often only after exhaustive efforts, and sometimes in exchange for proprietary technology. *Multinationals and the National Interest,* p. 71. See also Robert Z. Lawrence, "Japan's Low Levels of Inward Investment: The Role of Inhibitions on Acquisitions," *Transnational Corporations,* vol. 1, no. 3, December 1992, p. 47, who notes that, in contrast to most countries, new foreign investment in Japan occurs primarily through greenfield establishments and/or joint ventures. For an elaboration of this point, see Simon Reich, "Asymmetries in National Patterns of Foreign Direct Investment," p. 12. Citing an internal U.S. Treasury assessment, Reich concludes that "hostile takeovers are rare, and foreign takeovers usually occur only after all domestic possibilities have been exhausted," p. 13. See also Dennis Encarnation, *Rivals Beyond Trade: America versus Japan in Global Trade,* Cornell University Press, Ithaca, N.Y., 1992, and Mark Mason, *American Multinationals and Japan: The Political Economy of Japanese Capital Controls,* Harvard University Press, Cambridge, Mass., 1992.

unchanged, notwithstanding efforts by the Japanese government to address the problem.[266] In a 1995 White Paper, the Japan External Trade Organization (JETRO) noted that "whereas Japan has 15.4 times as much cumulative direct investment abroad as foreign investment at home, U.S. external investment is just 1.2 times the foreign investment within the (U.S.)...and the [comparable] figure for Britain is 1.3."[267] A recent U.S. government report supports this assessment, noting that "Japan's stock of inward foreign direct investment (FDI), relative to the overall Japanese economy, remains miniscule compared with that of other industrialized countries."[268]

The report also notes that Japan's outward investment flows dwarf investment into Japan, with a ratio of around ten to one throughout the 1990s. This contrasts with the investment flows of other OECD countries, which are much more balanced, with the ratio of outward to inward investment below two in every case—substantially below in most cases. These asymmetries continue. "In 1994, Japanese overseas FDI was $41 billion," while Japan's inward FDI was $4.2 billion.[269] This lack of receptivity to foreign investment on the part of Japan is considered a major trade barrier.[270]

If Japan stands out among the industrialized countries for its barriers to foreign direct investment, it is by no means alone in having barriers. As

---

[266] USTR, *1996 U.S. National Trade Estimate Report on Foreign Trade Barriers,* pp. 195–196. The report states that Japan has acknowledged that its inward investment lags far behind that of other industrialized economies and notes that the government of Japan has established a high level council charged with promoting measures to improve the investment climate. In 1995, the United States and Japan signed an agreement which commits Japan to take additional actions to promote foreign direct investment.

[267] Ibid.

[268] Ibid. In 1991 the OECD estimated Japan's share in global inward foreign direct investment to be under 1.5 percent, as compared with the 36 percent of FDI hosted by the United States. The Office of Technology Assessment also found that foreign investment in Japan is by far the most restricted, with foreign affiliates controlling 0.9 percent of total assets in Japan compared with 20.4 percent in the United States. See *Multinationals and the National Interest,* Office of Technology Assessment, p. 79. See also Robert Z. Lawrence, "Japan's Low Levels of Inward Investment."

[269] In part, this imbalance in flows reflects Japan's large trade surpluses, although others would argue, also in part, that the trade surpluses are to some extent the result of investment barriers. See the presentation of L. Chimerine to the conference *Sources of International Friction and Cooperation in High-Technology Development and Trade.* Chimerine suggests that while macroeconomic factors, such as the low U.S. savings rate, unquestionably play a role in the U.S. trade deficit, the causality can run both ways. He notes, for example, that Japan has trade imbalances with countries with both high and low savings rates as well as high and low budget deficits. See also Robert A. Blecker, *Beyond Twin Deficits,* p. 1. This view contrasts with that of many economists who maintain that the excess of domestic investment over domestic savings, plus some accounting identities, fully explains the trade deficit. Blecker argues that "poor trade performance or low foreign demand can cause *both* a widening trade deficit *and* a low national savings rate (including a high fiscal deficit), as well as the reverse." Ibid.

[270] USTR, *1996 U.S. National Trade Estimate Report on Foreign Trade Barriers,* p. 195.

noted, in a limited number of sectors, the United States restricts investment to some degree. In France, heavy state participation in the economy, including direct equity investments by the central government in leading companies as well as indirect investment through state controlled banks, coupled with formal and informal investment screening in strategic sectors, combines to create substantial barriers to freedom of entry in high-technology sectors. In Germany, extensive cross-holdings among banks and major corporate groups, with representatives of national banks serving on the boards of companies to which they provide loans, make acquisition of existing companies problematic.[271] China welcomes foreign investment critical to China's economic development plans, though often with strict conditions concerning technology transfer (see below), and prohibits a wide range of other foreign investment, especially in the service sector. And in newly industrializing Korea, there are fewer foreign multinational investments than in almost any other late-industrializing country.[272]

The explanations for this policy divergence among the industrialized countries, as well as those rapidly industrializing, are many and varied. In some cases the divergence reflects different traditions of corporate governance, especially in Japan and, to a lesser extent, Germany.[273] In the case of Japan, there are "considerably more regulations on business than [in] most other countries, and this undoubtedly obstructs the entry of new firms, both domestic and foreign, into the market. Many foreign firms, which are able to enter other markets, face greater difficulties in entering the Japanese market due to [government] regulations and administrative guidance."[274] In other

---

[271] *Multinationals and the National Interest,* p. 59.

[272] For a discussion of current Chinese policy, see *1996 Foreign Trade Barriers,* USTR, p. 56. For Korea, see Alice Amsden, *Asia's Next Giant,* p. 147. See also Robert Lawrence, "Japan's Low Levels of Inward Investment," p. 53. With only 18 percent of its FDI investment foreign majority-owned, Korea joins India (14 percent) and Japan (34 percent), "as conspicuous outliers," cited in Simon Reich, "Asymmetries in National Patterns of Foreign Direct Investment."

[273] Sylvia Ostry, "Technology Issues in the International Trading System." p. 15. The author observes that bank-centered governance arrangements are more resistant to mergers and acquisitions than are equity-oriented arrangements. Since equity markets are much more accessible through mergers or acquisitions, the result is significant asymmetry in the ability of companies to acquire new technologies and obtain market access.

[274] The Report of the Ad-Hoc Committee on Foreign Direct Investment in Japan, Keidanren Committee on International Industrial Cooperation, Committee on Foreign Affiliated Corporations, entitled *Improvement of the Investment Climate and Promotion of Foreign Direct Investment into Japan,* p. 5, cited in *Multinationals and the U.S. National Interest,* p. 76. See also Simon Reich, "Asymmetries in National Patterns of Foreign Direct Investment," p. 11 and *passim.* In recent years, analysts and managers of U.S.-based multinationals have argued that official government restrictions have been supplanted by private sector impediments. The USTR's 1996 Foreign Trade Barriers report notes that the current low level of FDI in Japan

cases, it reflects a tradition of a broader, more direct role for the state, particularly with respect to multinational companies.[275]

The effect of these "structural impediments" is to make foreign direct investment difficult, even exceptional, especially for the acquisition of existing firms in leading high-technology sectors. The administrative difficulty can provide, at a minimum, the opportunity to oppose unofficial performance requirements of the type pursued by European governments.[276] These obstacles also permit the exclusion of direct foreign investment in strategic sectors.[277] Indeed, in the case of Japan, the time and cost associated with overcoming the reluctance of the national administration and the constraints on mergers and acquisitions mean that many high-technology companies are either excluded or obliged to settle for licensing agreements.[278,279]

INVESTMENT BARRIERS, LICENSING AGREEMENTS, AND TECHNOLOGY TRANSFER

Licensing agreements, however, limit market access for firms with competitive products, particularly in Japan. Licensing is designed to ensure that the host-country firms gain access to the advanced technology of their foreign competitors. This is one explanation for the large, nonreciprocal flows

---

reflects years of exclusionary business practices, high market-entry costs, discriminatory bureaucratic practices, and the cumulative effect of years of formal restrictions on inward investment. P. 195.

[275] Carnoy et al., *The New Global Economy*, p. 91. For example, in 1985, forty-one of the world's largest 200 industrial enterprises were state owned (of these, eighteen were multinationals, of which seven were French). These included French flagship companies such as Elf-Aquitaine, Renault, CGE, St. Gobain, and Thompson.

[276] See Ostry, "Technology Issues in the International Trading System," p. 12, Laura Tyson, *Who's Bashing Whom?* and *Multinationals and the U.S. Technology Base.*

[277] Martin Carnoy notes that "government policies are crucial in determining how locally based MNEs are protected from foreign competition," with policies ranging from direct protection (i.e., reserving telecommunications or energy markets for domestic companies) to government procurement and ideological protection through antiforeign competition propaganda. Carnoy et al., *The New Global Economy.*

[278] Some acquisitions do take place in Japan, but they have been generally confined to domestic firms. See *Multinationals and the National Interest,* pp. 74–75. Citing Japanese Fair Trade Commission sources, the report notes that "in contrast to the limited number of merger and acquisition activity by foreign investors in Japan, such activity among domestic Japanese firms is vibrant and unhindered." Ibid., p. 74.

[279] Reich points out that determined foreign investors essentially have two options: greenfield site construction or licensing. Simon Reich, "Asymmetries in National Patterns of Foreign Direct Investment," p. 14. However, the high cost of land, unfavorable exchange rates, and regulatory roadblocks make the greenfield option available to only a few companies. The result is licensing and/or joint ventures which frequently involve the transfer of technology, leading to what Reich describes as "widescale, non-reciprocated technology transfers from the United States to Japan." Ibid., p. 15.

of technology from American to Japanese firms.[280]   This outflow of technology from U.S. and European firms to Japan is compounded by the technology transfer requirements of many rapidly industrializing countries.[281] Meeting these performance requirements can be a sine qua non for investment and even sales of advanced products.   As noted above, in strategic sectors such as aerospace, telecommunications, and automobiles, the imposition of compulsory technology transfer continues to grow.

In the United States, the relative openness of the investment environment became a source of growing concern in the latter half of the 1980s, a period of very rapid growth in direct foreign investment.[282]   For example, the rapid acquisition of U.S.-based suppliers of semiconductor equipment and materials led the U.S. semiconductor consortium SEMATECH to oppose—unsuccessfully—the transfer of these suppliers of strategic inputs to foreign ownership.[283]   The acquisition of companies and technologies of direct relevance to U.S. national security led to the Exon-Florio provision in the Omnibus Trade and Competitiveness Act of 1988. This provision empowers an interagency group, the Committee on Foreign Investment in the United States (the CFIUS) to veto the takeover of U.S. companies on national security grounds.[284]   However, this authority has been narrowly interpreted by the Treasury, and the CFIUS process has been criticized for its reactive, case-

---

[280] Ibid.

[281] See the section Compulsory Technology Transfers above. While many countries seek to upgrade their technological capability, some analysts argue that China, "because it enjoys tight governmental control of a huge potential market, has taken the practice to a new level." See the article by Blustein and Mufson, *The Washington Post,* 19 May 1996, citing Greg Mastel.

[282] In 1989, for the first time, foreign multinationals invested more in the United States than U.S. corporations invested abroad. Much of this inward investment flow was concentrated on the acquisition of U.S. high-technology companies. For example, from October 1988 to March 1993, 722 such companies were acquired, with Japanese investors accounting for 437, the United Kingdom 82, France 49, and Germany 29. The acquisitions of Japanese investors included 108 computer firms, 52 in semiconductors, 42 in materials, 36 in electronics, and 34 in semiconductor equipment. See Proctor P. Reid and Alan Schriesheim (eds.), *Foreign Participation in U.S. Research and Development: Asset or Liability?* National Academy Press, Washington, D.C., 1996, p. 80. See also Robin Gaster and Clyde V. Prestowitz, Jr., *Shrinking the Atlantic: Europe and the American Economy.* North Atlantic Research, Inc. and Economic Strategy Institute, Washington, D.C., June 1994.

[283] Howell et al., *Creating Advantage,* p. xiii. See also Andrew A. Procassini, *Competitors in Alliance: Industry Associations, Global Rivalries and Business-Government Relations,* Quorum Books, Westport, Conn., 1995.

[284] First created in 1975 by President Ford and formalized in the 1988 Omnibus Trade and Competitiveness Act, the CFIUS is composed of officials from the Departments of State, Commerce, Defense, Justice, and Treasury; the Office of the United States Trade Representative; the Office of Management and Budget; and the Council of Economic Advisers. In the 1988 Act, the president was granted the authority to investigate and block foreign investments that threaten national security.

specific approach.[285] Moreover, the CFIUS currently lacks the authority or resources to assess broad concerns about foreign investment in technologically strategic industries of great relevance to the defense base.[286]

FOREIGN DIRECT INVESTMENT IN HIGH-TECHNOLOGY INDUSTRY

U.S. policy has long held that foreign direct investment brings benefits in the form of jobs, capital, technology, tacit knowledge, and choice and price advantages for consumers. This view reflects the technological and commercial preeminence of the U.S. economy in the postwar world which sought to insure overseas investment opportunities for U.S. companies as well as the traditional American policy goal of maximizing consumer welfare. In the case of high-technology industry, there may be occasions when more nuanced views are required.

Given the dynamic, learning-by-doing characteristics of leading high-technology industries, some informed critics argue that broad generalizations are inadequate, especially as a guide for national policy. They point out that in certain specific circumstances, foreign investment can also have significant disadvantages for the host country.[287] For example, existing companies, and the benefits they furnish to the local economy, can be displaced by the arrival of foreign companies with significant competitive

---

[285] In 1990, the report by the Defense Science Board, *Foreign Ownership and Control of U.S. Industry,* recommended improvement of the CFIUS review process. In her 1992 analysis of high-technology trade, Laura Tyson raised the same concern, describing the CFIUS as "remarkably passive" in the exercise of its responsibilities. Laura Tyson, *Who's Bashing Whom?* p. 147. She notes that in some 700 notifications of investment, 600 involved acquisitions of foreign high-technology firms. Fourteen of these were investigated and one, the Chinese acquisition of an aerospace components manufacturer, was blocked. Ibid. The CFIUS does not have the authority to require foreign investors to notify the Committee of proposed investments, nor does the law require review of foreign investments in nonpublicly traded companies. U.S. General Accounting Office, *Foreign Investment: Analyzing National Security Concerns,* GAO/NSIAD-90-94, Washington, D.C., 1990.

[286] For a useful discussion of the CFIUS, see Reid and Schriesheim (eds.), *Foreign Participation in U.S. Research and Development.* This recent study also notes the limitations of U.S. government data concerning foreign acquisition of U.S. high-technology firms, and calls for the development of "more sophisticated capabilities for assessing and addressing the risks and capitalizing on the opportunities presented by the growth of foreign involvement in the nation's dual-use technology base." Ibid., pp. 78–79.

[287] Laura Tyson, *Who's Bashing Whom?* pp. 42–45. See also Alan Tonelson, who attacks what he calls "a techno-globalist assumption," i.e., that inward foreign direct investment is *always* good for the recipient; therefore any attempts to restrict or channel it are necessarily foolish. Tonelson points out that while foreign investment can add to the recipient country's wealth and to its wealth-creating capabilities, for a phenomenon as large and complex as foreign direct investment, broad generalizations about its effects offer little policy-relevant

advantages conferred by the national policies of the investing company's home government, e.g., subsidies or protected home markets.[288] The presence of well-financed, technologically advanced foreign companies can also deter the expansion of domestic companies or the entry of domestically based competitors through aggressive pricing made possible by more integrated corporate structures, more patient stockholders, and a willingness to tolerate losses in pursuit of a long-term strategy. In these conditions, foreign investment may actually reduce market competition, through either the elimination or purchase of domestic firms. This can result in more concentrated market power at both the national and global level.[289] In some cases, the outright purchase of relatively small companies providing key technology components for industries, such as semiconductors, can increase national dependency on foreign-owned suppliers while also augmenting the possibility of anticompetitive behavior.[290]

---

guidance, especially for high-technology competition. Alan Tonelson, "The Perils of Techno-Globalism," *Issues in Science and Technology,* Summer 1995. The OTA analysis cited above observes that greenfield investments in facilities for manufacturing, design, and R&D add more to an economy than does an investment in distribution networks, whose main effect is to make imports available, often to the detriment of local producers of such products (although to the benefit of consumers). The OTA analysis suggests that because "FDI can be concentrated in different sectors and deployed to very different effects...different forms of FDI can and do have different implications for the U.S. technology base." *Multinationals and the U.S. Technology Base,* p. 150 and chap. 6. *The need for nuanced assessment does not imply that FDI is not generally beneficial to the recipient economy.* See, for example, P. Reid and A. Schriesheim (eds.), *Foreign Participation in U.S. Research and Development,* pp. 66–80. For an analysis highlighting the benefits of foreign direct investment to the U.S. economy, in particular technology inflows and industrial "best practice," see Richard Florida, *International Investment: Neglected Engine of the Global Economy,* American Enterprise Institute, Washington, D.C., 1994. For a provocative discussion of U.S. policy on foreign direct investment, see Robin Gaster, "Guiding Foreign Investment," *Foreign Policy,* Fall 1992.

[288] This is especially true when the investing companies benefit from the direct support of their home government through equity investment, debt forgiveness or favorable loans by state controlled banks. See Laura Tyson, *Who's Bashing Whom?* p. 43. These conditions also suggest that market considerations are not dominant. "When a foreign government is the investor, however, foreign national interests are at play, and these may conflict with American national interests." Ibid.

[289] Ibid., p. 146. Much of the concern over foreign direct investment focused on the possible adverse effects of acquisitions in the U.S. semiconductor industry on market competition and national security. "Because semiconductors are an indispensable input throughout the electronics complex, strategic control over their supply by concentrated Japanese oligopoly poses a threat to downstream producers throughout the world." Ibid. See also Linda H. Spencer, *Foreign Investment in the United States: Unencumbered Access,* Economic Strategy Institute, Washington, D.C., 1991, and U.S. Department of Commerce, *Foreign Direct Investment in the United States,* Washington, D.C., 1993.

[290] Laura Tyson, *Who's Bashing Whom?* p. 146. For example, industry observers note that a sudden shortage of epoxy resin for semiconductor packaging (due in this case to an accident

CONSEQUENCES FOR COMPETITION POLICY
AND FOREIGN POLICY

This dependency can have significant consequences in terms of both national competitiveness and national security.[291] For example, a number of studies indicate that American companies relying on foreign suppliers for crucial components, such as DRAMs and displays, have been put at a competitive disadvantage because they cannot obtain the latest technology in the narrow timeframes that characterize competition in high-technology products.[292] Failure to obtain critical components in a timely manner can seriously disadvantage producers dependent on foreign supply. Indeed, geographically concentrated supply of key components, along with the relevant tacit knowledge, assigns suppliers of these vital inputs considerable strategic leverage. When supply is concentrated among a small number of producers, or a single country, to a significant extent suppliers can determine access to relevant technologies, the rate at which they can be incorporated into new products, and the price their customers must pay.[293]

Technological dependency heightens the risk of slower product evolution and of lower revenues to fund marketing and R&D for the next generation of products. More fundamentally, producers who depend on innovations in a distant foreign economy may have reduced opportunities to profit from

---

destroying a key plant) caused considerable concern among U.S. semiconductor producers who are dependent on foreign suppliers. In addition to epoxy for plastic packages, the U.S. industry relies on a limited number of Japanese suppliers for high-purity silicon, ceramic packages, and, to a lesser extent, the steppers which imprint circuits on semiconductors. See John P. Stern, "Japan: the Philosophy of Government Support for Information Technology," p. 3 and *passim.*

[291] A recent National Academy of Engineering report notes that the United States, much more than any other industrialized country, has relied on technology niche companies for a disproportionate share of major product and process innovation, adding that from the mid-1980s to early 1990s foreign acquisitions of these firms increased significantly. See P. Reid and A. Schriesheim (eds.), *Foreign Participation in U.S. Research and Development,* p. 70.

[292] Michael Borrus and Jeffrey A. Hart, "Display's the Thing," p. 24. Delays in access can have significant consequences. For a new electronics product, a six months' delay to market can cost up to one-third of its potential revenue stream. See George Stalk and Thomas M. Hoot, *Competing Against Time,* Free Press, New York, 1990. Studies documenting the delay in the delivery of key components, such as semiconductor manufacturing equipment to American producers by Japanese suppliers, include the Defense Science Board Task Force's *Foreign Ownership and Control of U.S. Industry,* June 1990; the National Advisory Committee on Semiconductors' report *Preserving the Vital Base: America's Semiconductor Equipment and Materials Industry,* 1990; and the General Accounting Office's *U.S. Business Access to Certain Foreign State of the Art Technology,* GAS/NSIAD-91-278, September 1991. European users encountered similar problems. See also Howell et al., *Creating Advantage,* pp. 116–132, especially pp. 129–130.

[293] Borrus and Hart, "Display's the Thing," p. 24.

spillovers and localized tacit knowledge.[294]   In some cases this can be overcome, a prospect aided by the rapid advance in electronic communication.   However, the sustained efforts to acquire and apply new foreign technologies are more easily undertaken by large multinationals in the most advanced countries.   Such efforts are more difficult for small and medium-size firms.[295]   On the other hand, the success of newly industrializing countries such as Korea and Taiwan—particularly in electronics—suggests that industrial groups with effective government support can overcome locational disadvantage.   The presence of multiple, competing suppliers of equipment embodying the most recent technology is a crucial determinant of their success.[296]

In addition to the competitive risks of technological dependency on concentrated foreign suppliers, the transfer of control over strategic military technologies through foreign purchase of U.S.-based suppliers of key components raises important national security issues.[297]   When foreign governments are the investors or when an acquisition threatens to concentrate market power among few suppliers, some analysts believe careful review by the CFIUS is warranted to determine if the acquisition should be blocked or subject to conditions.[298]   As noted above in the section on dual-use technology, a key feature of America's defense strategy is superiority in high-technology weapons systems.   Concentrated market power in the hands of

---

[294] For example, Jonathan Eaton and Samuel Kortum find that distance appears to inhibit the flow of ideas between countries, although trade relationships enhance such flows. Also, increased levels of education significantly facilitate a country's ability to adopt technology. See J. Eaton and S. Kortum "Trade in Ideas: Patenting and Productivity in the OECD," *Journal of International Economics,* 1996 forthcoming.

[295] Borrus and Hart, "Display's the Thing," p. 24, note 5.

[296] SEMATECH's development of advanced equipment technologies for semiconductor manufacture contributed importantly to the development of the Korean semiconductor industry by generating alternative sources of supply equipment for manufacturing, which led to a more competitive market for DRAMs. In addition, some U.S. firms reportedly transferred DRAM designs to Korean firms, presumably to ensure alternative sources of supply.

[297] See the report of the Defense Science Board Task Force *Foreign Ownership and Control of U.S. Industry.* The report cites a number of instances where assured access to military technologies has been threatened by foreign acquisition of U.S. firms and the subsequent transfer of the government-supported research offshore. Pp. 15–19. The report also underscores the relationship between national security and the long-term economic competitiveness of U.S. industry. P. 14. The "dual-use" strategy described by the Under Secretary of Defense for Acquisitions, Paul Kaminski, explicitly relies on commercial technology as the leading agent of change in technology areas critical to U.S. defense capabilities. See the presentation by Paul Kaminski in C. Wessner (ed.), *Sources of International Friction and Cooperation in High-Technology Development and Trade.*

[298] Laura Tyson, *Who's Bashing Whom?* p. 44. The conditions imposed on Hüls' acquisition of MEMC is an example. See the contribution of Charles Cook, MEMC vice president for strategy and acquisitions to the NRC symposium, in *International Access to National Technology Promotion Programs, 19 January 1995 Symposium Proceedings.*

foreign producers, even close allies, can significantly reduce national autonomy in the implementation of foreign policy goals.[299]

The lack of consensus which characterizes post-Cold War foreign policy initiatives creates uncertainty about assured supply of components for military systems. For example, foreign producers may not be willing to supply the components to be utilized in military actions, either as a result of home government "guidance" or through the reluctance of foreign firms to disrupt supply schedules to meet military surge requirements.[300] National companies, no matter how multinational in operation, are responsive to the policy preferences of their home governments.[301] And foreign policy goals deemed central in one polity may have considerably less resonance in another, especially if compliance involves financial loss or market disruption.

Because technological dependence poses substantial risks to national autonomy, whether the national goal is competitive markets or the ability to project military force in support of national interests, the state of the national technology "supply base" is a legitimate concern of government. In the case of the United States, policymakers should recognize that national technological competency is a central concern of U.S. competitors. Consequently, national policies designed to maintain national technological capabilities may become as normal a part of U.S. policymaking as they are for most other governments, with specific attention directed to foreign direct investments.[302]

---

[299] Jeffrey A. Garten, *A Cold Peace: America, Japan, Germany, and the Struggle for Supremacy*, Times Books, New York, 1992, p. 227. The author points out that U.S. trading partners could exercise real leverage over the United States as creditors, as suppliers of vital technology, or as hard-to-please partners in endeavors that Washington finds important.

[300] For example, a recent NAE report cites a 1983 case where the Japanese government reportedly pressured the leading Japanese producer of ceramic materials, Kyocera, to stop supplying ceramic nose cones to the U.S. Tomahawk Missile program through its U.S. subsidiary. See P. Reid and A. Schriesheim (eds.), *Foreign Participation in U.S. Research and Development,* p. 77.

[301] Carnoy et al., *The New Global Economy in the Information Age,* p. 84. Citing Stephen Cohen, the author points out the unwillingness of U.S. corporations to cooperate in the development of the French independent nuclear deterrent, or to construct the Soviet-European pipeline. Formal export controls on technologies, such as U.S. restrictions on the Soviet purchase of modern telecommunications equipment (e.g., fiber-optic cable), or the stated reluctance of major Japanese producers to supply flat panel displays to the U.S. military, or the difficulty in acquiring key components during the Gulf War, illustrate both the risks of dependency on foreign supply and the responsiveness of corporations to the policies of their home government.

[302] *Multinationals and the U.S. Technology Base,* pp. 150–153. The report identifies five basic types of Foreign Direct Investment in ascending order of their contribution to the U.S. technology base: distribution facilities for imported products; final assembly facilities for imported components; manufacturing facilities that use a mix of imported and locally manufactured components; integrated design, engineering, and manufacturing facilities that provide customized products for the local market; and fully integrated research and production facilities that are a strong strategic component of a firm's global R&D, sourcing, and manufacturing operations. P. 150.

Greater awareness of the strategic importance of certain core technologies for the national economy does not suggest that borders should be closed to investors or that quality investment should not be encouraged. On the contrary, for some technologies this policy focus might suggest foreign investment should be actively encouraged.

In the absence of domestic production, foreign investment has distinct advantages. However desirable indigenous production and autonomous technological capabilities are as national goals, foreign direct investment, especially high–value-added investment, is distinctly superior to dependence on imports from foreign suppliers who locate their production, employment, and research facilities in their home markets.[303] European policymakers have adopted this approach for some time. In order to mitigate the effects of dependency on foreign electronics manufacturers, European policymakers have steadily increased local-content requirements for existing electronics facilities while instituting generous promotional measures to attract high–value-added investment by global electronics companies.[304] The promotional measures have included R&D subsidies and regional development subsidies tied to European production locations.[305] **As noted above, a multilateral agreement should be pursued in order to limit the level of investment incentives, as well as protectionist measures, designed to attract international investment flows.**

CONSEQUENCES FOR HIGH-TECHNOLOGY COMPETITION

It is important to recognize that the current differences in investment regimes acquire great significance in the context of global high-technology competition. Substantial advantage can be obtained through the acquisition of new, technology-rich, capital-poor companies or through assured access to important markets. Conversely, competitive advantages can be denied through

---

[303] Tyson, *Who's Bashing Whom?* p. 251. Richard Florida, *International Investment*, emphasizes the relationship between international investment and productivity, particularly as a result of transplant manufacturing facilities, which can generate a general rise of productivity in domestic manufacturing through a virtuous cycle of imitation, adaptation, and improvement. p. 21.

[304] Tyson, *Who's Bashing Whom?* p. 251.

[305] Ibid. Laura Tyson notes "the tendency of EC member states to outbid one another in subsidies to attract inward investment," a practice the EU Competition Commission is attempting to regulate. See the section on investment incentives above. For a generally positive assessment of EU efforts to ensure "quality" investment, see R. Gaster, "Guiding Foreign Direct Investment," pp. 97–100. Gaster notes that in addition to traditional measures such as tariffs, quotas, and antidumping measures, the EU has set anticircumvention standards to prevent firms from escaping trade barriers or dumping duties by exporting nearly finished products into European "screwdriver" factories for final assembly.

limiting access to important segments of global markets, or through the requirement of disproportionate investments of management time and capital for establishment. Restrictive policies are especially important—and distorting— when they result in the systematic transfer of technological assets to competing firms based in the country applying restrictive practices.[306]

SANCTUARY MARKETS

For companies competing in the rapidly changing high-technology markets, the consistent inability of firms to invest in the home economies of their major competitors represents a profound disadvantage. With much of global trade accounted for by intrafirm transactions, the exclusion of competing firms from a major segment of the global market is a significant disadvantage for firms competing from contestable domestic markets.[307] Put simply, if one set of national competitors has access to three-thirds of the relevant technologies and world markets, and competitors from other nations have access to two-thirds of the technology and markets, over time, the competitive position of the second group of firms is likely to deteriorate.[308]

Because the majority of U.S. trade with Japan takes place within affiliated networks of Japanese firms with investments in the United States, this has major implications for the current and future U.S.-Japan bilateral trade deficit. It may also explain in part the lack of adjustment in the trade deficit despite major movements in currency values.[309] This situation contrasts with the U.S. trade and investment relationship with Europe. Transatlantic investment and trade are more balanced and have proven far more

---

[306] Simon Reich, "Asymmetries in National Patterns of Foreign Direct Investment," *passim* and pp. 32–36.

[307] The importance of investment for trade flows was reviewed extensively in *Multinationals and the U.S. Technology Base,* p. 2 and chap. 6. The study found that affiliates of foreign-based multinationals account for a substantial portion of U.S. merchandise trade and the greatest share of the merchandise trade deficit. While the U.S.-European direct investment relationship has been relatively symmetrical, the U.S.-Japan investment relationship is asymmetrical. Japanese investment in the United States exceeds U.S. investment in Japan by a three-to-one ratio, and is concentrated more in wholesale operations than in manufacturing, as compared with European and American investment patterns. P. 3.

[308] Borrus and Hart, "Display's the Thing," p. 50.

[309] Simon Reich, "Asymmetries in National Patterns of Foreign Direct Investment," p. 32. It is important to note that quite recently the Japanese surplus has declined. Some of this change may be due in part to the liberalization of the Japanese economy; much of it is due to the movement of Japanese production offshore. The decline in the trade surplus with the United States is correlated with a rapid rise in U.S. imports from other Asian countries (e.g., electronics from China) that, in many cases, are produced with Japanese capital and technology.

responsive to macroeconomic forces than have U.S. investment and trade patterns with Japan.[310]

Moreover, the imbalance in investment opportunity translates into quite different competitive environments. Companies operating in markets open to foreign investment face greater competition than do firms which benefit from relatively closed investment regimes. Competition among domestic firms in protected markets may occur; though this restricted competitive environment lends itself to cartel behavior. Moreover, the ability to develop new technologies and products, test the market, and maintain higher domestic prices for products than those found in more competitive export markets, all work to the advantage of the firms benefiting from a sanctuary market. The higher rents thus obtained are particularly important for product development in capital-intensive industries.[311]

BILATERAL SOLUTIONS?

Offsetting the competitive advantage offered by markets effectively closed to foreign investors, and therefore trade opportunities, will of necessity remain a prime objective of competitors for market share in high-technology products. The U.S.-Japan Semiconductor Agreement was designed to address this problem and seems to have offered an effective solution, both for the U.S. and Japanese industries and for their governments. Because of its importance for competition in the semiconductor industry, and its potential application to other high-technology trade conflicts, the conditions leading up to the Semiconductor Agreement and its unique features are discussed in some detail in Supplement A (below).

While it has many detractors, often on grounds of principle rather than effect, the Semiconductor Agreement represents a policy success in that it resolved the problem it was designed to address.[312] Foreign market share in Japan increased, not only to the benefit of American producers (and Japanese users), but also for newly competitive Korean and Taiwanese produc-

---

[310] *Multinationals and the U.S. Technology Base,* pp. 14–16. See also Robin Gaster and Clyde V. Prestowitz, Jr., *Shrinking the Atlantic.*

[311] Simon Reich, "Asymmetries in National Patterns of Foreign Direct Investment," pp. 33–34 and *Multinationals and the U.S. Technology Base,* pp. 12–16 and 153.

[312] Laura Tyson summarizes traditional economists' views of import agreements, such as the STA, as resting "on three *a priori* arguments: that (they)...result in the cartelization of markets; that they increase prices by limiting competition; and that, by violating the nondiscrimination principle of the GATT, they undermine the world trading regime. All three of these arguments are subject to qualification on analytical grounds alone." She adds that none of these expectations was confirmed by the actual experience of the Semiconductor Agreement. Tyson, *Who's Bashing Whom?* p. 134.

ers. The fundamental virtue of the agreement is that it provided a mechanism to open the Japanese market rather than closing the American market. It also provided a measurable target for access and a monitoring mechanism to promptly address dumping in the United States or third-country markets. The Agreement also set the stage for much-improved relations between the U.S. and Japanese semiconductor industries.[313] Initially as a result of MITI encouragement, the industry also witnessed a proliferation of design-in projects between foreign suppliers and Japanese users. Subsequently, these relations expanded exponentially in response to market demands with an explosion of joint ventures and other long-term alliances between Japanese and foreign companies.[314] After a difficult start, the Agreement also largely removed, or at least contained, a major source of tension within the overall U.S.-Japan bilateral relationship.

The successful conclusion of the August 1996 negotiations on the renewal of a modified Semiconductor Agreement between Japan and the United States underscores the desire of both countries to avoid renewed trade conflict in this strategic sector. While the Japanese government was initially reluctant to renew the government-to-government agreement, a compromise was reached which included two separate but interdependent agreements: an agreement between the U.S. and Japanese industry associations on international cooperation in the semiconductor sector and a related agreement between the U.S. and Japanese governments. (See Box I for a summary of the agreements). The inter-industry agreement encourages continued industry-to-industry cooperation and includes the monitoring of shares in key mar-

---

[313] For example, Japanese government sources point out that design-ins, in which foreign semiconductor devices are specifically designed for inclusion in commercial products (such as camcorders or VCRs), increased nearly ninefold in Japan since the first agreement was concluded in 1986. See Ministry of Foreign Affairs, Japan, *Salient Points and Data Related to the Japan-U.S. Semiconductor Arrangement*, p. 4. Reflecting the title, the Ministry's report states that, "the Arrangement's historical role has been fully achieved." Despite this accomplishment, the report does not recommend continuing the agreement, a view which corresponds closely with a January 1996 Electronics Industry Association of Japan (EIAJ) report, *Mission Accomplished: Why There Is No Need for a Semiconductor Agreement with Japan.* Somewhat to the surprise of the American government, in early 1996 the Japanese government rebuffed the initial U.S. effort to renew a modified agreement, i.e., one which would not have a specific target figure. See Susumu Maejima, "Ending Chip Pact Is Shortsighted...," *Asahi Evening News,* 5 February 1996.

[314] As examples, the EIAJ cites the case of Toshiba and Motorola, which have a long-standing joint venture, established in 1987, in Sendai in northern Japan. More recently, Hitachi and Texas Instruments agreed on a new joint venture in 1995, located in Richardson, Texas, although their collaboration dates back to 1988. EIAJ, *Salient Points.* For a valuable industry perspective on the considerations relevant to joint ventures, see the presentation by Motorola's Charles White in C. Wessner (ed.), *Sources of International Friction and Cooperation in High-Technology Development and Trade.*

BOX I. THE U.S.-JAPAN SEMICONDUCTOR
AGREEMENTS OF 1996.

The United States and Japan faced the possibility of renewed conflict in the semiconductor sector as they approached the scheduled expiration of the 1991 U.S.-Japan Semiconductor Agreement on 31 July 1996. The U.S. semiconductor industry and government, while acknowledging that tremendous progress had been made in increasing U.S.-Japan industry cooperation and foreign market access in Japan under the 1991 Agreement and its 1986 predecessor, were seeking a renewal of the Agreement to ensure continuation of the government-to-government and industry-to-industry mechanisms that had led to the progress of the previous decade. The Japanese industry and government argued that, inasmuch as the foreign share in the Japanese market had reached more than 30 percent by the first quarter of 1996, the objectives of the earlier agreements had been fulfilled and no new agreement was necessary.

In particular, while the U.S. government and industry wanted a continued governmental role in monitoring progress in market access and industry cooperation, the Japanese government resisted any continued government role. The 1991 Agreement provided that the two governments jointly calculate foreign market share in Japan on a quarterly basis and set a 20 percent foreign share target against which to measure progress in increasing foreign market share in Japan.

After months of negotiations between both the two industries and the two governments, a compromise was reached on 2 August, avoiding renewed conflict in this sector and permitting continued industry-to-industry cooperation with a continued, but reduced level of government monitoring. Key to this compromise was an agreement that the two industries, rather than the two governments, would calculate foreign shares in key markets around the world, and then report this information to the two governments. In addition, no new market share target was set. These provisions, which met both sides' objectives, were included in two separate but interdependent agreements: an agreement between the U.S. and Japanese industry associations on international cooperation in the semiconductor sector and an agreement between the U.S. and Japanese governments concerning semiconductors.

Inter-Industry Agreement

The inter-industry agreement provides for a continuation of the cooperative activities between semiconductor users and suppliers begun under the 1986 and 1991 Agreements. A joint industry steering committee is to continue as under the previous agreements to review developments in joint industry cooperative activities to promote the design-in of foreign semiconductors in new products in specific sectors and in emerging applications. In addition, the agreement provides for new cooperation between semiconductor suppliers in areas such as standardization; environment, worker health and safety; intellectual property rights; trade and investment liberalization; and market development. The new agreement also creates a Semiconductor Council made up of senior executives of the U.S. and Japanese industries, which is to meet once a year.

The agreement also provides that industry experts are to prepare quarterly market/trade flow reports, which are to include data on market shares of foreign semiconductor products in major semiconductor markets. The reports are to be based on a number of data points, largely from industry surveys and government data, allowing the two industries jointly—or, if necessary, each acting independently—to measure the foreign share of the Japanese and other markets. These quarterly reports are to be

continued

provided to the governments for their review, thereby ensuring a government role in monitoring developments in foreign market access in Japan.

While the initial inter-industry agreement is bilateral, a provision is included that is to permit the European and other semiconductor industries to join the Semiconductor Council once their respective governments have eliminated or agreed to eliminate expeditiously all tariffs on semiconductors. No term is set on the inter-industry agreement, although the two industry associations agreed to review the agreement in three years.

Government-to-Government Agreement

The government-to-government agreement welcomes the inter-industry agreement and affirms the intention of both the U.S. and Japanese governments to support industry-to-industry cooperative efforts. It further provides that the two governments are to hold consultations at least once a year to receive and review the reports on data collected by the industries; to review and discuss the cooperative activities conducted under the inter-industry agreement and market trends and developments, including those relating to competitiveness and foreign participation in major markets; and to discuss government policies and activities affecting the semiconductor industry.

The U.S. and Japanese governments also agree to establish a Global Governmental Forum made up of governments of major semiconductor-producing countries and other interested economies. This Forum is to meet annually to discuss issues that may affect the global semiconductor industry. The organization is to be open to governments without any precondition, and its first meeting it to be held no later than 1 January 1997.

The two governments also reaffirm in the agreement the need to avoid the problem of injurious dumping through effective and expeditious antidumping measures, consistent with the GATT and the WTO Antidumping Agreement. The term of the government-to-government agreement is three years, expiring on 31 July 1999.

kets, but without a new foreign market share target for the Japanese market. Interestingly, while the inter-industry agreement is bilateral, it provides for the possibility of participation by other semiconductor industries on the condition that their national tariffs on semiconductors are eliminated. The government-to-government agreement affirms the support of both governments for cooperation by their respective industries and provides for regular consultations on the evolution of the semiconductor industry and policies affecting it.

The success of the Semiconductor Agreements to date and the continued difficulty in obtaining market access for other high-technology products has led some analysts to conclude that similar agreements offer a pragmatic and effective means of resolving long-standing trade conflicts.[315] In this view,

---

[315] For one of the most complete statements of the rationale for this approach, see Michael Borrus et al., *The Highest Stakes,* pp. 197–205. See also Stephen Cohen and Pei-Hsiung Chin, "Tipping the Balance." For an industry view, see Council on Competitiveness, *Roadmap for*

arguing that U.S. trade partners should change their domestic policies—or their enforcement—is less likely to produce desirable results, for either party, than is a focus on reciprocal access to markets, investment opportunities and technologies.[316] Whatever the merits of this view, in the absence of significant advance in a multilateral context, pressure for bilateral solutions focused on specific outcomes is likely to grow.

Whatever solutions are proposed—multilateral talks, bilateral initiatives to reduce structural impediments, or regional cooperation—the sharp differences in national investment regimes are an important source of friction and a destabilizing element in the global system. As discussed above in the section on cooperation, unequal access for trade and investment undermines the basis for sustainable international cooperation in the development of new technologies. These differences in access, or asymmetries, are also a major source of trade imbalances, and generate pressures for restrictions on foreign investment in countries which do have relatively open investment regimes. Finding a means to address these asymmetries is therefore an important instrument for both improved cooperation and reduced trade friction. Progress in reducing these asymmetries in national practices requires a two-track approach: (1) a determined and sustained effort to improve access on a bilateral basis and (2) the conclusion of a workable multilateral accord on investment.**

Reflecting the importance of investment issues for the continued health of the international trading system, a multilateral effort to reach an international investment accord is now under way. In May 1995, members of the OECD decided to launch negotiations among the twenty-five OECD member countries, with the aim of reaching a Multilateral Agreement on Investment (MAI) by mid-1997.[317] The objective of the negotiations is to "provide a broad multilateral framework for international investment." This involves both a broad definition of investment and a broad coverage of

---

Results, p. 55. For a discussion of the advantage of a results-oriented approach in terms of Japanese perceptions and practices, see James Fallows, *Looking at the Sun,* pp. 448–450 and 497–498, and Laura Tyson, *Who's Bashing Whom?* pp. 133–136 and chap. 7. For a succinct review of the strategic trade theory underlying this approach, see Luc Soete, "Technology Policy and the International Trading System: Where Do We Stand?" pp. 3–7. For an informative discussion of views of past and current "strategic trade" theorists, see Jeffrey Hart and Aseem Prakash, "Implications of Strategic Trade for the World Economic Order."

[316] In some cases, bilateral solutions may be the only recourse possible, not least because barriers to market entry can include collusive private arrangements, informal government guidance, formal regulations obstructing access to investment, discriminatory standards, and issues of competition policy and its enforcement, none of which are under the purview of the WTO or any other multilateral mechanism.

[317] See the June 1995 OECD Ministerial Communiqué. SG/PRESS (95) 41. Paris, 24 May 1995.

sectors, levels of government, and types of laws and regulations. The goal of the OECD negotiations is to achieve "high standards for the liberalization of investment regimes and investment protection," including effective dispute settlement procedures, with limited general exceptions, such as national security.[318] This is the major reason for the decision to first negotiate among OECD members rather than among the members of the more broadly based WTO, though it is not clear that this strategy will prove successful.[319] The ultimate goal of the negotiations is a "free standing international treaty" that, though negotiated at the OECD, will not be confined to OECD member countries.

If an *effective* multilateral accord on investment, such as that envisaged in the OECD negotiations, can be obtained, it would represent a major step forward on investment issues. For the United States, it is essential to continue to actively pursue this multilateral approach, while seeking to extend to nonmembers acceptance of the principles and standards to be agreed at the OECD. At the same time, systematic review of the impact of foreign acquisitions on the U.S. defense base and related consequences for the U.S. competitive position should be conducted.[320] This is especially important in oligopolistic industries where foreign competitors have a history of anticompetitive behavior or where the foreign investor is, in effect, a foreign government.[321] Indeed, in neither case can the interests of the participants in the transaction be considered paramount.[322] It is therefore appropriate that sustained attention be given to the nature and sectoral concentration of inward foreign investment, a practice already adopted by many governments.

---

[318] See the presentation "Progress toward a Multilateral Agreement on Investment" by William Witherell in Peter S. Rashish, (ed.), *Building Blocks for Transatlantic Economic Area.* European Institute, Washington, D.C., 1996. Witherell describes a wide range of mechanisms available to the OECD for consultations. Ibid., p. 32.

[319] Some analysts believe this decision to have been an error, because non-OECD members are skeptical of negotiations to which they are not party, though as noted the OECD is to undertake consultations with interested nonmembers. For a critical assessment of the OECD initiative, see E. M. Graham, presentation to the Institute for International Economics conference on the *World Trading System: Challenges Ahead,* 24–25 June 1996.

[320] See Laura Tyson, *Who's Bashing Whom?* pp. 145–149. The author was quite critical of the operation of the CFIUS. In her 1992 analysis, she argues that "a prudent policy toward foreign direct investment when it involves a product or industry deemed critical to national security should follow two basic principles. First, it should use requirements for national ownership or local production by foreign suppliers to enhance national control over suppliers regardless of their nationality. Second, it should seek a diversity of suppliers to maintain a competitive global supply base." Ibid., p. 147. At that time, the author faulted the CFIUS for having ignored both the potential national security threat and the strategic economic threat posed by foreign oligopolistic control over critical industries such as semiconductor producers and its major equipment and materials suppliers. Ibid., p. 148.

[321] See the section Foreign Direct Investment in High-Technology Industry above.

[322] Laura Tyson, *Who's Bashing Whom?* p. 43

For example, many countries have implemented investment policies designed to maintain advanced manufacturing and the associated high-wage employment.[323] Given the importance of high-technology industry for the U.S. defense base and future economic growth, an enhanced awareness of other countries' policies toward foreign investment could prove valuable to U.S. policymakers.[324] Moreover, in light of the role of the United States in the world economy, a better understanding of inward foreign investment in host countries abroad as well as in the United States, and policy decisions taken on the basis of that information might well contribute to the success of ongoing negotiations on a multilateral investment agreement. Better informed decision-making offers the prospect of strengthening the U.S. technology base and improving opportunities for high-wage employment.

COMPETITION POLICY—CONVERGENCE?

Market concentration in high-technology industries poses special challenges to national policymakers. The dangers of oligopolistic control in high-technology industries are not well-understood. As noted above, national policies and instruments to defend domestic industries from predatory pricing or concentrated market power can encounter substantial opposition both on grounds of principle and as a result of their impact on other commercial interests. Antidumping duties are often seen as an imperfect and incomplete response to the conditions—effective closure of the national markets of major exporters, despite commitments to liberalize—that give rise to injurious dumping. However, existing international rules do not adequately address the market barriers and competition policy failures that make antidumping remedies necessary.

Ultimately, what is needed is a network of enforceable commitments by trading nations to have *and to enforce* laws against private restraints of trade. The actual convergence of those national laws, beyond the core anticompetitive practices which they all must prohibit, may take considerable time. To the

---

[323] See R. Gaster, "Guiding Foreign Investment," pp. 97–100. As a result of these policies, "foreign firms in Europe have been forced to raise the local content of their finished products." Gaster also notes that European politicians have pressured foreign firms to produce sophisticated goods, train local workers, introduce advanced management practices and "explicitly called on foreign firms to conduct more R&D in Europe and encourage them to set up design facilities and use local suppliers, training the suppliers to meet international standards." Ibid., p. 98. Robert Scott notes the European concern with maintaining employment in manufacturing and the policies adopted to encourage high-wage employment in high-value manufacturing. See Scott,"Trade: A Strategy for the 21st Century" in Todd Schafer and Jeff Faux (eds.), *Reclaiming Prosperity: A Blueprint for Progressive Economic Reform.* M.E. Sharpe, New York, 1996, pp. 249–250. For a proactive policy approach, see the Commission *Green Paper on innovation,* p. 23.

[324] A recent National Academy of Engineering study recommends the U.S. government improve its capability to assess and track foreign investment. See P. Reid and A. Schriesheim (eds.), *Foreign Participation in U.S. Research and Development,* p. 78.

extent convergence is eschewed or delayed, however, there will be potential for friction, and pressure for other policy remedies will grow.

In addition, there is increasing recognition that concern over oligopolistic behavior in high-technology industries should be extended "to include control over strategic assets and capabilities."[325] As electronics pervade the modern economy, industrial innovation depends on the components, materials, automated machinery, and control technologies (i.e., software) that are combined to create new products and processes. A domestic economy's effective access to those technological inputs depends crucially on the national supply base.

CUMULATIVE CONSEQUENCES

The supply base affects producers in several ways. A robust supply base provides access to relevant technologies at affordable costs and in a timely fashion while offering opportunities for essential interaction between suppliers and producers. The quality of the supply base can therefore be a crucial determinant of domestic firms' ability to compete in rapidly evolving high-technology products.[326]

However, the flows of technical know-how across national borders are determined by the distinct national characters of local markets. For example, markets in the United States tend to be relatively open, employee mobility is high, foreign firms can acquire technologically advanced American firms, and the capital constraints of U.S. firms often encourage the licensing of core technologies.[327] This contrasts with East Asian countries such as Japan or Korea where markets are less open, the mobility of skilled labor is low, direct acquisition of firms is virtually impossible, integrated industrial structures (in keiretsu or chaebol) provide firms with patient capital, and national institutions are less open to foreign access.[328] The result of these structural differences is that accrued technological know-how tends to be retained locally and tends not to diffuse rapidly across national boundaries.[329]

---

[325] Laura Tyson, *Who's Bashing Whom?* p. 275.

[326] Dependence on distant producers of key components poses unique competitive challenges. Perhaps the most important concerns the nature of the dependency. Some companies outsource to expand capacity and thereby leverage their own knowledge. Other firms rely on outsourcing for knowledge, that is, the company needs a component but lacks the in-house skill to produce it. The second form of dependency leaves the firm open to competitive challenge. Charles Fine and Daniel Whitney, "Is the Make-Buy Decision a Core Competence?" pp. 17–18. It is assumed that companies cannot avoid some degree of dependence; the point here is that the form this dependency takes has quite different competitive consequences. Ibid., p. 21.

[327] Borrus et al, *The Highest Stakes,* p. 22.

[328] Ibid.

[329] Ibid.

GLOBAL COMPETITION
AMONG NATIONAL COMPANIES

There is a paradox in that as economic activity expands across the globe, the role of national policies and practices continues to play a major role, sometimes even determining competitive outcomes. The globalization of economic activity is, however, very real.[330] At the risk of underscoring the obvious, many high-technology enterprises are multinational in *operation,* performing research, manufacturing, assembly, and sales at many sites in many countries. And, as noted above, the number and variety of interfirm relationships continues to expand. For technology, they range from licensing agreements to joint development and acquisition. For manufacturing, they range from original equipment manufacturing, to second sourcing, to assembling and testing with similar diversity in marketing and service agreements as well as formal joint ventures to develop and manufacture new products. (See Box E.) These transnational relationships have become so numerous and pervasive that some observers have concluded that they portend the end of national commercial rivalry, to be supplanted by a system of global cooperation leading to enhanced global welfare.[331,332]

Without disputing the growth in the scope of economic competition, and

---

[330] As one analyst notes, "Texas Instruments' high-speed telecommunications chip is a model of the efforts of technology enterprises to marshall global resources in the design, manufacture, and delivery of a product. TI's TCM9055 chip was designed with Ericsson engineers in Nice, France, using software developed in Houston, Texas. The chip is produced in Japan and the U.S., tested in Taiwan, and included in Ericsson products that monitor systems in Australia, Mexico, Sweden, and the U.S. Chip-testing [for] assembly plants in the Philippines is done by engineers at terminals in Dallas. Marketing is organized into worldwide product teams rather than by country; technology glitches are solved by engineers from multiple countries communicating with each other from work stations around the globe." Daniel M. Price, *Investment Rules and High-Technology: Toward a Multilateral Agreement on Investment.* OECD, Paris, France, 26–27 October 1995, p. 4. The author concludes from this global activity that "while geographic location is important to customer service and product support, national borders appear increasingly irrelevant."

[331] Howell et al., *Creating Advantage,* p. xiv. This view is in fact often encouraged, for good reason, by foreign investors wishing to blend into local economies as good corporate citizens subject to the same treatment as domestic firms.

[332] While the reality of global economic activity is incontestable, its meaning is subject to different interpretations, with substantial relevance for national policy. For example, a recent Japanese industry publication states that, "given the borderless nature of many fundamental operations in this [the semiconductor] industry, the era in which it made sense to distinguish the nationality of a semiconductor chip has long since passed." Electronic Industry Association of Japan, *Mission Accomplished,* Tokyo, January 1996, p. 1. The recent Japanese Ministry of Foreign Affairs report makes a similar point, arguing that in light of the many alliances in the industry "the [semiconductor] Arrangement, which attempts to identify the nationality of semiconductors, is now obsolete and meaningless." February 1996, p. 1. The same report, however, then describes the market in terms of national shares.

therefore the scope of operations of global corporations, others argue that economic globalization "simply expands the dimensions and complexity of the competition being waged between national systems."[333] In this view, most multinational corporations, though not all, remain profoundly national in *character* even while operating on a global scale.[334] This national character is the result of a complex yet powerful combination of factors. First and foremost the corporations are subject to the norms and policies of the home nation. More pervasively, to the extent that board members, senior executives, and key employees are nationals of the same country, they tend to reflect its values, beliefs, culture, and historical experience. This gives meaning and effect to the norms and accentuates the importance of national policy, for example, on antitrust, bribery, or export controls. Corporations also depend on the national educational and technological infrastructure, as well as on the trade and investment policies of a given country and in some cases on specific government support.[335] From this perspective, far from being rendered irrelevant by the globalization of competitive functions, government policies can play an instrumental role in determining competitive outcomes in the rivalry between national enterprises competing on a global scale.[336]

NATIONAL STRATEGIES
FOR MULTILATERAL SOLUTIONS

Systematic differences in access to national markets for trade and investment pose serious risks for the international system. Mutual reductions of tariff barriers, while valuable in opening markets to trade, afford only part of the resolution of international trade and investment issues for high-tech-

---

[333] Howell et al., *Creating Advantage,* p. xiv.

[334] There are, of course, cultural variations—which underscores the basic point. U.S. and British multinationals, more than any other country's, are philosophically antistate and therefore tend to be more "footloose." Although distinctly "American" or "British" in modes of operation (e.g., employment patterns or local sourcing, and especially when in need of government help in securing economic interests) "they are first and foremost profit-making private enterprises, obligated to their...stockholders, not to any nation-state." Carnoy et al., *The New Global Economy in the Information Age,* p. 86.

[335] Martin Carnoy argues that Japanese and certain European multinationals may not be state-owned but they tend to be "state-tied," that is, their operations may be shaped by state policies that allow them to gain advantage over competing firms. More generally, he argues that large international companies "are rooted in national economies and national systems of production" as they operate through affiliates in other countries. Carnoy et al., *The New Global Economy,* p. 87.

[336] Howell et al., *Creating Advantage,* p. xiv. Michael Porter also supports this view. He suggests that "differences in national economic structures and institutions contribute profoundly to competitive success." Furthermore, he stresses that as a result of the globalization of competition, the role of the home nation is more rather than less important. *The Competitive Advantage of Nations,* p. 19.

nology industries. Intervention by government in high-technology industry is on the increase, as is the number of countries employing these practices, with each country seeking to alter the composition of its national economy to achieve a competitive edge.[337] In addition, the United States, which continues to be the leader and the initiator of proposals to improve the international economic system, must do so in the face of a decline in its relative weight in the world economy, the special concerns of European policymakers for their high-technology sectors, and the markedly different investment and trade practices of some East Asian economies. **All these factors suggest that if further progress is to be achieved, a coherent, long-term strategy will be required.**

At the outset, this report recognized the inevitability of friction as countries seek to acquire and maintain the advantages of high-technology industry in an increasingly global economy. This "competitive friction" can, as noted above, have its virtues insofar as it stimulates human endeavor and productive national investment. Yet, friction also has its risks. It can easily degenerate into conflict. Multilateral agreements and mechanisms to eliminate practices which generate friction, and contain or mitigate it when it occurs, are essential. However, they may not be sufficient; indeed, they are not likely to be reached if current trends in the targeting of high-technology industry continue. And these practices are likely to continue to the extent that they are seen as successful and relatively costless. Sustained attention at the national level, supported by institutions and policies designed to address these issues, will prove necessary to contain friction and help give positive direction to the competition among nations.

In fact, international friction over these practices need not be inevitable. Where either a country has avoided the competitive challenge by exiting the industry (e.g., the United States in color televisions) or where the industries involved do not recognize or do not accept the challenge (e.g., automobiles in the United States in the 1970s and perhaps today in Europe), friction can be avoided or at least delayed. However, the absence of friction and conflict does not indicate the absence of harm to national interests. In economies open to trade, the industrial base can be eroded and future development compromised, through both the loss of productive capability in high-growth industries and the loss of related research and development activity, thereby narrowing future technological and political choices.

Some level of friction may therefore be positive. For example, the process of granting access to national technology programs in Europe and the United States must be informed by the realities of access to foreign technology and markets through both trade and investment. To gain reciprocal access to technologies, export markets, direct investment opportunities, and

---

[337] See OECD, *Industrial Subsidies* and Box B above.

the balanced trade they engender, a pragmatic approach, adjusted to the special conditions of high-technology competition, is required. In the case of the United States, a pragmatic approach might also contribute to moderating the moralistic tone that sometimes characterizes U.S. commentary on countries whose positions or practices differ from those of the United States, while helping to assure that the United States has the economic strength to negotiate effectively and provide essential global leadership. In the case of Europe, structural reform to facilitate the creation and ease the operation of small innovative firms could contribute substantially to economic growth and employment.[338] Indeed, the continued growth of the advanced economies, and the corresponding ability of the governments in Europe and the United States to maintain their unparalleled research infrastructure and liberal trading policies, are key conditions for sustained, global economic growth and the progress this implies for mankind.

To meet this challenge, the consistent application of result-oriented policies offers the prospect of opening markets for investment and trade for all participants in the global economy, and of curbing abuse in areas essential to industrial innovation, such as intellectual property protection. Sustained attention on the part of major participants in the international system to the special characteristics of high-technology trade could contribute to strengthening the multilateral system.

The major participants in the international trading system already have policies designed to improve the attractiveness of their national economies to high-technology industry in sectors such as microelectronics and aerospace. In some cases, further efforts to improve the regulatory environment in sectors such as telecommunications and biotechnology could offer substantial benefits to national economies. Investment in infrastructure, especially in universities and other research institutions, is a form of competition which is likely to enhance both national and global welfare. In contrast, national policies which, over a sustained period, emulate the corporate strategy of always being a "fast second," presume that countries with well-developed research infrastructures both will have the economic means to sustain investment in these systems and will continue to have the necessarily public support to maintain open access. Similarly, roughly reciprocal access to markets, investment opportunities, and enabling technologies appears fundamental to the continued health of the multilateral system, as does some international understanding on the appropriate rules-of-the-game for the support of targeted industries. To achieve these goals, countries are likely to pursue both bilateral and multilateral agreements, an approach that holds out the prospect of the incremental success necessary to the maintenance of the system as a whole.

---

[338] See European Community, *Green Paper on innovation,* pp. 5–7.

Main Points of the Summary Report

THE IMPACT OF COMPETITION FOR NEW TECHNOLOGIES AND NEW INDUSTRIES ON INTERNATIONAL RELATIONS

For the reasons outlined in this report, the competition for high-technology industry shows every sign of intensifying as governments seek to capture the benefits of economic growth, high-wage jobs, and political autonomy for their citizens and of the "political space" successful programs offer governments. Moreover, in the absence of the bonds provided by the Cold War need for a common defense, the political saliency of these competing policies and programs and the tensions they engender will become an increasingly important element of international relations. Insofar as the political tensions created by these national programs are unlikely to diminish, it is important that they receive sustained attention from policymakers.

Improved understanding of the rationale for government support for high-technology industry and the recognition that governments will continue to actively support high-technology industry within their borders are essential for effective policymaking. An open discussion of the effectiveness of these programs, including their effects on the multilateral system, could provide the basis for reducing the friction these programs generate. Better documentation of the goals of such programs and of the nature of the measures designed to achieve them can make the terms of the current competition among nations more transparent, and thereby improve the possibility of international agreement on appropriate means and limits for such programs. Developing "rules of the game" for programs supporting high-technology industry cannot, however, be seen as a substitute for open and contestable markets for these industries. Effective national and international competition within a transparent policy framework is an important means of encouraging the rapid development of new products and processes.

RECOGNIZING THAT DIFFERENT ASSUMPTIONS AND GOALS MEAN DIFFERENT POLICIES

Reducing international friction resulting from governments' efforts to promote national industries requires that full recognition be given to the differences in national objectives and values. Many countries do not seek to maximize short-term consumer welfare, although the assumption that they do often underlies discussions of Anglo-American policymaking. Instead, a growing number of countries seek to create national comparative advantage to capture the special benefits that high-technology industries impart to an economy and its workforce. Understanding that countries' objectives legitimately differ is an essential step in effective policymaking, both to advance national economic interests and to reduce international friction.

THE IMPACT OF HIGH-TECHNOLOGY COMPETITION ON THE STRUCTURE AND DISPOSITION OF RESEARCH

The increased emphasis on short-term research with a focus on commercially relevant applications is changing the modes of operation of major corporate R&D facilities, with the phase-out of large research laboratories and the reduction of the basic research effort of large companies. Similar pressures and concerns are changing the way universities and other research institutions fund, carry out, and publicize research. Universities and research institutes face challenges in terms of funding, mission, and relationship to national programs. The exploitation of commercially relevant university research, a concern accentuated by the financial pressures faced by universities, raises complex questions as to the appropriate disposition of results and associated benefits. Because of the importance of these developments, further work on the impact of changes in public and private support for basic research is required. Similarly, further work on the role of university and research institutes in national and international science and technology research should be undertaken.

NEW TECHNOLOGIES AS A SOURCE OF COOPERATIVE ALLIANCES

The race to exploit the opportunities inherent in new technologies generates powerful incentives for greater cooperation between otherwise competing companies and national programs. New product development increasingly involves companies in a broad array of complex technologies and production processes with high capital costs and special expertise, encouraging alliances across sectors and national boundaries. Some of these alliances are technology driven; others result from the actions of governments. In the latter case, these policies often explicitly target the acquisition of new technologies for the domestic economy. While the need for government intervention in private alliance activity is limited, it is equally important to recognize the role of government action in creating alliances and the impact of alliances on the competitive environment. In some cases, where government action is the driver of alliance activity, other governments may have legitimate concerns as to the goals, distribution of benefits, and competitive consequences. Further data collection and analysis in this area could improve understanding of the causes and consequences of different types of strategic alliances.

THE CHALLENGES POSED BY
INCREASED COOPERATION

The costs, complexity, and risk associated with the development of new technologies will continue to encourage greater international cooperation by public and private entities. As the scope and intensity of this international interaction increase, the potential for friction will rise as well. Further work should be undertaken on the problems and prospects of international cooperation, particularly in commercially relevant areas. It should focus on principles to guide international cooperation, incorporating the lessons of existing international programs for the organization of cooperative research; it should also recognize that such cooperation is a valuable option and an opportunity, not a right.

TECHNOLOGY DEVELOPMENT PROGRAMS
WILL REMAIN NATIONALLY BASED

Technology development programs will continue to be driven by national goals, whether for missions such as defense, energy, or the environment, or broader economic objectives of the nation. These motivations will continue to constrain, though not eliminate, opportunities for international cooperation. Successful international cooperation requires that the limitations of national objectives and other factors be taken into account.

These limiting factors include

• asymmetries in the structure and funding of national programs,
• the different technological competencies and assets nations or firms bring to a cooperative enterprise,
• the related perception that some countries are not contributing their "fair share" to basic research, and
• inadequate and ineffective intellectual property protection and investment regimes which discriminate against foreign acquisition and fail to provide national treatment (formally or informally).

The frictions generated by the asymmetries in national technology programs may become more acute as advanced high-technology companies based in countries with significantly less-developed research infrastructures seek to participate in national programs of the leading industrial countries.

FURTHERING COOPERATION THROUGH AN
INTERNATIONALLY AGREED NATIONAL BENEFITS TEST

Given the underlying differences in rationale, structure, funding, and accessibility of national programs, a formal international agreement "guaranteeing access" is unlikely. Less formal understandings may offer a means of assuring greater transparency, and ultimately result in greater foreign participation in government-funded civilian R&D programs. For example, a sustained effort to reduce conditionality, perhaps through the construction of an objective, internationally accepted national benefits test, might be undertaken on a multilateral basis. The advantage of a multilateral effort would be to

- make national performance requirements more transparent,
- establish agreed-upon guidelines, and
- focus on contributions to the national technology base rather than on corporate "nationality."

A sustained multilateral effort could also seek to improve understanding of differences among national technology development programs. For example, it could gather improved data concerning formal rules for participation in national or regional technology programs, supplemented by objective assessments of current administrative practices, i.e., actual foreign participation and its rationale, rather than theoretical "openness."

CONDITIONS FOR SUSTAINABLE
INTERNATIONAL COOPERATION

The growth and success of international cooperation will thus be determined by the terms and conditions of proposed cooperation as well as by the history of previous cooperation. The degree of agreement on shared priorities, equitable technical contributions (not merely financial contributions), and a shared capacity to exploit the results of cooperation will determine the willingness of both firms and public authorities to participate in international cooperative ventures. The perception of decisionmakers concerning their opportunities to exploit the results of cooperation is a key consideration. Consequently, the contestability of the end-product markets is likely to become an essential condition for productive and sustainable international cooperation.

THE LINKAGE BETWEEN INTERNATIONAL
TECHNOLOGY COOPERATION AND
A LIBERAL TRADE REGIME

Greater international cooperation in technology development will therefore be determined in part by developments in the multilateral trading system. An open, market driven, international trading system provides an environment conducive to sustainable international cooperation for the shared development of new technologies. Long-term cooperative efforts, and the cooperative spirit they presuppose, coexist with difficulty in an environment marked by trade disputes or inadequate respect for the explicit and implicit rules of the game.

Consequently, a key condition for sustained international cooperation in the development of new technologies is improved adherence to the principles of a liberal trade regime. Closed national markets, whether through quotas, discriminatory standards, biased public procurement, or private anticompetitive practices, undermine the political and policy conditions necessary for effective international cooperation. Reciprocal access to national technology development programs presumes equal access to end-use markets.

DOMESTIC POLICIES WITH INTERNATIONAL CONSEQUENCES

Efforts to further technological cooperation, particularly public/private cooperation, therefore imply parallel efforts to further trade liberalization in areas "within the borders," such as government procurement, national treatment for foreign investment, and effective competition policy. Sustainable cooperation implies a competitive, transparent procurement regime, the right of establishment for foreign investors, including roughly comparable regimes for the acquisition of existing firms as well as market access for final products resulting from such cooperation.

EFFECTIVE POLICYMAKING REQUIRES INTEGRATED INSTITUTIONAL STRUCTURES

Because there are powerful, reciprocal relationships between trade and technology policy, effective national policymaking requires institutions which reflect this relationship. Institutions which link trade and technology policy are most effective. Many countries have established national institutions with the capacity to assess, coordinate, and implement the various policies impacting the development of national high-technology industries. Countries such as the United States, which often do not follow an integrated approach to international competitiveness in high-technology industries, can impose needless costs on their consumers and producers while putting unnecessary stress on the international trading regime.

The absence of effective foresight or alternative policy mechanisms to support a promising industry has led some countries to impose protectionistic measures, often because few other policy tools were available to decisionmakers. And protectionist border measures, in the absence of a coherent policy framework, can impose significant costs on those domestic producers dependent on imports of foreign components, as well as on the consumers of final products.

CHALLENGES FOR THE WTO

The new World Trade Organization will face significant challenges in seeking to contain and adjudicate the disputes arising from fierce competition for high-technology industry. The legitimate desire of countries to encourage high-technology industry through a variety of largely domestic measures collides with the legitimate concern of other countries that these measures will disadvantage their national industry. For example, one country's program of technology diffusion can appear to its competitors as an effort to support domestic producers at the expense of foreign exporters. Similarly, policies to encourage technological innovation by national firms can be seen as an effort to create national advantage in international markets.

When "normal policies" of support for innovation and diffusion are supplemented by systematic restrictions on access to markets for trade and investment, bilateral policy approaches are likely to be adopted while multilateral solutions are sought. Pressures to seek trade agreements with measurable outcomes are likely to become more frequent. Indeed, in the short term they may be the only market-opening alternative available to policymakers. As the absence of investment opportunities generates greater pressure for negotiated outcomes concerning access to and shares of regional markets, the need for progress on an effective, enforceable investment accord becomes more compelling.

It is widely accepted that governments should support research. Similarly, government support for R&D for government functions is also widely accepted. There is less consensus on government support for applied research. Yet the distinctions between basic and applied research are often difficult to make and rarely decisive in defining the appropriate government role. Existing GATT definitions and exemptions for these categories of state aid involve considerable ambiguity. Definitional difficulties for the current exemptions for R&D and the environment are therefore likely to become a source of international controversy. International initiatives to refine definitions and advise on disputes are unlikely to prove satisfactory. Consequently, the decision to provide such an exemption should be revisited. This is not to suggest that government R&D subsidies are necessarily improper. But where such subsidies distort international trade and cause injury, they should remain actionable.

PROGRESS ON GOVERNMENT PROCUREMENT

Public procurement remains a major means of government support for national industries and a significant source of friction in the international system. Because a significant share of markets for high-technology products is derived from public purchases, discriminatory public procurement of high-technology products has sparked major trade disputes. Governments continue to see public procurement as a means of supporting national champions through noncompetitive contracts and government procurement decisions continue to have an important impact on trade flows.

A reexamination of the way the multilateral trading system addresses government procurement is now necessary. In the aftermath of the Uruguay Round agreement, government purchases are one of the few areas not covered in a thorough manner by international trade disciplines. To a large extent, this is because the existing Government Procurement Agreement (GPA) requires its signatories to make the leap to full national treatment in government procurement, a leap that most countries remain unwilling to make. The alternative—described in Supplement E—would include all WTO contracting parties as members of a new GPA and would adopt the GATT tariff reduction procedures as a model that could be applied to achieve steady market-access improvements in the government procurement area.

A broader, more effective agreement would also offer a means to reduce a major source of friction in high-technology trade while providing the benefits of transparent competition to government acquisitions of high-technology products. In addition, such an agreement would be a major step toward improving the transparency and due process in government procurement which would help reduce the impact of corruption on trade in high-technology products.

With respect to international cooperation, proponents of such cooperation must recognize that cooperation to develop new technologies implies transparent and competitive procurement regimes. Reserving markets for national champions is ultimately incompatible with sustained cooperation. This is especially true when firms benefiting from protected home markets seek access to the publicly financed technology development programs in countries with more open markets. Consequently, the contestability of participating firms' home markets may become a de facto condition for cooperative activity. A revised Government Procurement Agreement could thus reinforce efforts to encourage international cooperation.

MUTUAL RECOGNITION ON STANDARDS

Discriminatory or exclusionary standards practices are also incompatible with efforts to improve international cooperation in the development of new products. International cooperation is an excellent means to avoid conflict over differing national standards for key technologies. Recent calls for negotiations on a full and complete mutual recognition agreement for medical devices, telecommunications terminal equipment, information technology products, and electrical equipment, as well as a common registrations dossier for new drug products, should be supported.

INTELLECTUAL PROPERTY PROTECTION

There is a pressing need for sustained international commitment to the protection of intellectual property rights, which underpin much of the technological progress in sectors such as electronics and biotechnology. At the same time, a cooperative effort should be undertaken to adjust existing intellectual property rules to the needs of new fields such as global information systems and biotechnology. Multilateral efforts to constructively address these issues should be encouraged.

THE SPECIAL DYNAMICS OF COMPETITION IN HIGH-TECHNOLOGY INDUSTRY MEANS EFFECTIVE ANTIDUMPING POLICIES REMAIN NECESSARY

In technology industries characterized by scale and learning economies, forward pricing strategies can be indistinguishable from predatory pricing. The high cost, rapid innovation, and short product cycles characterizing these industries make it possible for significant damage to domestic industry to occur in relatively short periods. Moreover, the higher returns which accrue to national firms benefiting from these practices can provide the resources to fund additional research, more rapid product development, expanded marketing, and overseas acquisitions of competitors. Even when practiced for relatively short periods, these strategies provide substantial competitive advantage in high-technology markets. For the recipients of dumped products, the revenue losses from both reduced exports and reduced domestic market share are compounded by the loss of the dynamic efficiency gains, i.e., learning by doing, that characterize these industries. The cumulative effect of these practices can permanently alter the terms of international competition by forcing competing firms to exit a product market or by deterring new entrants.

In these circumstances, the need for prompt and effective antidumping policy at the national level is heightened. This may be a second-best policy solution, but it is likely to prove essential for countries with relatively open markets in high-technology goods. From the international perspective, unilateral national action could be usefully supplemented by improved consensus and standards on competition policy and its enforcement.

INVESTMENT POLICIES: THE NEED FOR A MULTILATERAL ACCORD

While the policymakers in some countries welcome the benefits of foreign direct investment and consequently have relatively open investment regimes, many other advanced industrial countries and rapidly industrializing countries do not share this view or apply the same open policies toward foreign investment. The explanations for this policy divergence among the industrialized countries, as well as those rapidly industrializing, are many and varied. However, the effect of these "structural impediments" is to make foreign direct investment difficult, even exceptional in these countries, especially for the acquisition of existing firms in leading high-technology sectors. As a result, many high-technology companies are obliged to settle for licensing agreements or find themselves excluded from important markets.

COMPULSORY LICENSING

Licensing agreements, especially compulsory agreements, may work to the serious detriment of the innovating firm. And restricted market access can have powerful negative effects on the competitive position of companies denied the economies of scale and other competitive benefits so important to high-technology industries.

INVESTMENT INCENTIVES

Better understanding of the scope of these incentives, their aggregate costs and their impact on decisionmaking at the level of the firm would be useful to national and regional policymakers. Within national jurisdictions it may be possible to establish norms—with legislative underpinnings if required—to limit excessive public-private transfers in the competition for new investment. At the international level this subject should be explored as a potential element for inclusion in the negotiations for a multilateral agreement on investment.

REDUCING NATIONAL DIFFERENCES
IN INVESTMENT ACCESS

The sharp differences in national investment regimes are an important source of friction and a destabilizing element in the global system. Unequal access undermines the basis for sustainable international cooperation in the development of new technologies. These asymmetries in national investment policy are a major source of trade imbalances and also generate pressures for restrictions on investment in countries which do have relatively open investment regimes. Finding a means to address these asymmetries is therefore an important instrument for both improved cooperation and reduced trade friction. Progress requires a two-track approach: (1) a determined and sustained effort to improve access on a bilateral basis and (2) the conclusion of a workable multilateral accord on investment.

In the case of the United States, the systematic collection of information and sustained policy level attention to direct foreign investment, particularly acquisitions, in high-technology sectors would improve understanding of their impact on the U.S. defense base and the U.S. competitive position. In the case of Japan and the industrializing East Asian economies, more open investment regimes would offer substantial benefits while also contributing to more balanced, and therefore more sustainable, trade flows. In the case of Europe, easing the costs and administrative burden involved in establishing new firms, as well as acquisitions of existing firms, would offer enhanced opportunities for economic growth and employment.

Supplements

SUPPLEMENT A.
HIGH-TECHNOLOGY COMPETITION IN SEMICONDUCTORS

The history of the semiconductor industry beginning in the late 1970s illustrates several aspects of international competition in high-technology industries.[339] The stakes of this high-technology competition were enormous. It is not widely recognized, for example, that the U.S. electronics industry, including semiconductors, is larger than the U.S. steel, automobile, and aerospace industries, combined. Employment in the semiconductor industry and related industries is equally significant. U.S. semiconductor makers employ nearly 240,000 people in the United States alone and semiconductors are the enabling technology for the nearly $400 billion U.S. electronics industry, which employs approximately 2.5 million Americans. In addition, the manufacturing capability of the industries mentioned above and, therefore, their global competitiveness depend directly on timely access to new, electronic manufacturing technologies. Moreover, because the strategic significance of the industry is widely recognized by policymakers around the world, "the semiconductor industry has never been free of the visible hand of government intervention."[340]

A Catch-Up Strategy

From its position as a late industrializer, the Japanese government recognized the strategic importance of semiconductors to its economy, in terms both of economic growth and national autonomy. In the 1970s Japan made a systematic effort, under MITI guidance, to promote a domestic semiconductor industry.[341] Building upon their experience in consumer electronics,

---

[339] This section draws heavily from the contributions to Committee deliberations of George M. Scalise, *The Nature of High-Technology Competition,* 13 December 1996 and William J. Spencer, *Technology (Transfer) at SEMATECH.* See also the excellent analysis presented by Nortel's Claudine Simson, Samsung's Y.S. Kim, and Motorola's Owen Williams in the discussion led by BRIE's Michael Borrus to the conference *Sources of International Friction and Cooperation in High-Technology Development and Trade, 30–31 May 1995.*

[340] Laura Tyson, *Who's Bashing Whom?* p. 85. For a comprehensive review of the extensive and ubiquitous government programs designed to develop and support the technologies underpinning the semiconductor industry, see Thomas Howell et al., *Creating Advantage, passim.*

[341] See the partial list of joint research and development projects in microelectronics sponsored by MITI in Laura Tyson, *Who's Bashing Whom?* p. 96. Many of these programs continued through the 1980s. In addition to the VSLI manufacturing program, MITI focused on promising technologies such as optical semiconductors, high-speed computing devices,

the large Japanese electronics firms invested heavily in the development of semiconductor technology. By the early 1980s Japanese producers had become a major force in the market for dynamic random access memories (DRAMs).[342]

Escalating Trade Friction

The vertically integrated structure of Japanese industry provided major advantages with respect to the capital-intensive investments required for state-of-the-art semiconductor facilities. In this high-tech competition, the Japanese companies benefited from their unique system of corporate governance, which places considerably less emphasis on quarterly or annual profits than does the U.S. system. Moreover, the rich revenue stream derived from previous success in penetrating the consumer electronics industry permitted cross-product subsidization within companies. At the time, a lower cost of capital encouraged the major Japanese electronics firms to make major counter-cyclical investments.

These structural advantages enabled the Japanese industry to undertake a massive capacity build-up in the early 1980s. The Japanese firms then accelerated their gains in market share through highly aggressive price cutting. The U.S. industry sustained progressively heavier losses through each generation of DRAM. By 1984–85 with the introduction of the 256K DRAM, the market share of the U.S. industry had gone from roughly ninety percent of the DRAM market in the late 1970s to less than ten percent, with most U.S.

---

synchrotron lithography, and optoelectronic devices, among others. Japan is not alone in providing R&D support in this sector. In the United States, the Defense Advanced Research Projects Agency (DARPA) has extensive technology development programs, though these are generally small scale and focused on "over-the-horizon" technology development, with an emphasis on defense requirements.

[342] This strategy was not without its costs. While direct central government expenditure was limited, though important, these programs presumably involved costs in terms of foregone consumer welfare or opportunity costs imposed on other industries, as a 1988 study by Baldwin and Krugman sought to determine. However, this concern for consumer welfare, however laudable, reflects values and assumptions characteristic of American-trained economists, which are not universally shared. See James Fallows, *Looking at the Sun*, chap. 4, especially pp. 180–190, and the section Producer-Oriented Economies above.

As Laura Tyson notes, "it is important to emphasize that the objective of the Japanese industrial policy in the semiconductor industry, as in other industries targeted for development, *was not consumer welfare*." (Pp. 86–87, italics added.) Tyson also notes that the Baldwin and Krugman study did conclude that the Japanese policies "succeeded in gaining a share of the 16K DRAM market for Japanese producers, but only at the expense of a net reduction in Japanese economic welfare." Laura Tyson queries the import of their findings because of their failure to take into account the possible dynamic effects of a Japanese position in the growing DRAM market on profits and technological externalities in future generations of semiconductor products and in systems incorporating semiconductor technology.

firms exiting the DRAM market entirely. A similar though less severe trend occurred in erasable programmable read-only memories (EPROM). The drastic effects of the Japanese competition led many informed U.S. observers to question the future viability of the U.S. semiconductor industry.[343]

The Japanese challenge to the U.S. industry was based on three interlocking elements. Foremost among these was the quality of Japanese production processes and products. The second, which helped drive the first, was the higher rates of investment made possible by the more integrated, capital-rich structure of the Japanese companies. A third key advantage enjoyed by the Japanese industry in comparison with the U.S. industry was its ability to exclude its global competitors from access to its domestic market. The formidable barriers to investment and the lack of market access for foreign producers of semiconductors provided a crucial benefit in an otherwise global industry.[344] It gave the Japanese industry a sanctuary, from which they could insulate themselves from competitive counterattacks and preserve a financial cushion for future product development. This enabled Japanese producers, operating from the closed Japanese market, to dump product on the global marketplace. Because of the shortness of the semiconductor product cycles and the speed with which the foreign markets responded to dumping, the impact of dumping on their U.S. competitors was pronounced.

The U.S. Industry Response
to the Japanese Trade Challenge

Leading figures in the U.S. semiconductor industry increasingly recognized the unique features of the Japanese competitive challenge. The combination of a closed Japanese domestic market, the dumping encouraged by the excess capacity in microelectronics products (encouraged by the capital-rich Japanese keiretsu), and the prospect of supply dependency for key components with the accompanying vulnerability to manipulation of price and availability posed a competitive threat that they were unlikely to surmount independently.

To meet this threat to the future viability of their industry, several U.S. companies joined in the mid-1980s in filing dumping actions against the Japanese semiconductor producers with respect to 64K DRAMs and EPROMs. In 1985 the Department of Commerce self-initiated a case in reaction to

---

[343] Laura Tyson provides an excellent analysis of the competition for dominance in the semiconductor industry. See Laura Tyson, *Who's Bashing Whom?* chap. 4, "Managing Trade and Competition in the Semiconductor Industry," pp. 85–113, as does the recent study by Kenneth Flamm, *Mismanaged Trade?* especially chap. 3, 4, and 5.

[344] Tyson, p. 87.

Japanese dumping of DRAMs of 256K and above. During roughly this same period, the Semiconductor Industry Association brought an action against the Japanese semiconductor producers under Section 301 of the Trade Act of 1974 to address the closed nature of the Japanese semiconductor market.

The dumping cases resulted in findings of substantial dumping and reasonable indications of injury. The remedy from these cases normally would have been high antidumping duties imposed on the products at issue. However, the U.S. semiconductor industry viewed such a remedy as both inadequate and flawed.

Antidumping duties were considered inadequate because before they could be imposed, most U.S. producers could no longer justify investing in new DRAM capacity and were forced to leave the DRAM market. No compensation would be provided to the U.S. industry; the effect of the antidumping duties would be prospective, that is, only to offset future dumping. And antidumping duties on a previous- or current-generation product was little deterrent to the dumping of subsequent products, especially if over-capacity and low marginal costs added to an incentive to capture market share by dumping.

Equally important, antidumping duties were a flawed remedy for an industry such as semiconductors. Given the resulting rapid decline in U.S. DRAM production, antidumping duties would impose a burden on the U.S. customers of the semiconductor industry, either by raising the price of DRAMs to them or by forcing them to shift to overseas production of their electronic systems products.

To meet this policy challenge, the American producers adopted a novel approach, ultimately accepted by the American government.[345] This approach did not seek to protect domestic producers of semiconductors from global competition nor to insulate the U.S. market from technological advance, which would ultimately disadvantage their U.S. customers. Rather than close the U.S. market, the U.S. producers sought to ensure access to the Japanese market for U.S. and foreign suppliers.

The U.S. approach sought not only to bring an end to dumping and a return to cost-based competition in the U.S. and other foreign markets, but also to remedy the problem of effective market access in Japan. Previous efforts to obtain equivalent access to the Japanese market had proved elu-

---

[345] While the Semiconductor Agreement and SEMATECH represent innovations in U.S. trade and technology policy, the Reagan administration arguably had a much more pragmatic and innovative approach to economic policy issues than is generally recognized, even when compared with the announced "results-oriented" trade policies of the Clinton administration. See, for example, Alan Tonelson, "Beating Back Predatory Trade," *Foreign Affairs,* July–August 1994. Tonelson argues that the efforts to support the semiconductor industry were but one element of a series of successful policies to rescue and restore beleaguered U.S. industries such as steel, machine tools, textiles, and automobiles.

sive. No matter how market conditions changed or how well U.S. products competed elsewhere in the world, the U.S. market share in Japan generally stayed around 10 percent,[346] well under U.S. shares in other markets. Efforts such as the formation in 1982 of an informal group, the U.S.-Japan working group on high-technology industries, did not prove successful. Although a series of recommendations by the group was adopted in 1983, by mid-1985 the U.S. share of the Japanese market was lower than it had been when the recommendations were first adopted.

Recognizing the failure of this consultative approach and of the Japanese government to fulfill undertakings on market access, the U.S. industry filed a Section 301 action. Sanctions could be authorized against Japan under a Section 301 action for failure to live up to its agreements or for engaging in unjustifiable and restrictive trade practices. For the U.S. industry, however, the objective was to obtain effective market access, not to penalize the Japanese.

The Semiconductor Trade Agreement

In 1986, just prior to the conclusion of the various legal proceedings, the U.S. industry agreed to suspend the dumping cases, and an agreement was reached on the market access issue. The settlement took the form of a five-year accord called the Semiconductor Trade Agreement (STA). Importantly, the Agreement included a side letter in which the Japanese government recognized the fact that the U.S. industry expected the *foreign* share of their market (not only U.S. producers) to grow and to exceed 20 percent by the year 1991.[347] The Agreement and the policies associated with its implementation included three principal features:

• Provisions designed to lead to concrete market access for non-Japanese firms,

• A specific target acknowledged as a threshold for foreign market share, and

• A mechanism to monitor semiconductor cost so that dumping could be addressed quickly and effectively with the prospect of sanctions should the terms of this agreement not be respected.

---

[346] As noted above, given the enormous R&D costs associated with the semiconductor industry and the resulting importance of competing in all markets to recoup these costs, access to the Japanese market, now the largest in the world, is considered essential for competitiveness in this global industry. Council on Competitiveness, *Roadmap for Results,* p. 55.

[347] The existence of the side letter was first revealed with the publication of Clyde Prestowitz, *Trading Places,* Basic Books, New York, 1988. For a discussion of the negotiations, see Kenneth Flamm, *Mismanaged Trade?* chap. 4. For a discussion of the impact of the side letter agreement specifying 20 percent as a reasonable expectation for foreign market share by 1991, see Kenneth Flamm, *Mismanaged Trade?* pp. 279–293.

An End to Dumping

The metrics of success for the STA were clear-cut: assured access to the Japanese market for foreign producers and an end to dumping. There would be no offsetting penalties for dumping and no restricted access to the U.S. market. Instead, dumping would cease and an effective deterrent mechanism was devised under which each individual producer would monitor its own cost data to detect dumping. For market access, an objective target for a minimum foreign market share was established at 20 percent of Japan's annual semiconductor consumption. This was a target that focused on access in fact rather than in theory. Because market penetration could be measured, it provided a concrete measure of success. This represented a significant departure from traditional U.S. process-oriented trade policy to a more results-oriented approach to market access; it also provided discipline.

Through Government Commitments

To be effective, the STA needed government commitment. If there was to be no U.S. government response to noncompliance—the common perception in U.S. industry of the U.S. government's behavior with respect to past trade agreements with Japan—there would be little prospect of the STA's meeting its objectives. After widespread evidence of noncompliance with the STA became available in late 1986, President Reagan imposed sanctions in April 1987.[348] Following this action dumping in world markets ceased, and the market share in Japan of U.S. and other foreign producers began to climb. By 1990 the increased share of the Japanese market made possible by the agreement translated into additional sales of over $1 billion for American companies alone.[349]

And Industry Cooperation

The success of the STA also depended heavily upon industry commitment. The STA not only established a realistic target for market access, but

---

[348] A Council on Competitiveness study concluded that a credible government commitment to sanctions may be necessary to ensure compliance with trade agreements, especially when they affect the interests of powerful foreign companies. The report notes that, "in the case of semiconductors, the U.S. government's commitment to sanctions was not credible until they were actually applied in 1987. Japanese industry and government were reportedly very surprised by the decision." Council on Competitiveness, *Roadmap for Results,* p. 56. The report adds that the U.S. government's commitment to sanctions contributed to the subsequent success in winning an extension of the Trade Agreement in 1992 and to the dramatic increase in the purchase of U.S. semiconductors by Japanese users.

[349] Laura Tyson, *Who's Bashing Whom?* p. 113. These additional sales also meant significant increases in R&D, capital investment, and high-wage employment.

was accompanied by major efforts by both U.S. and Japanese industries to pursue the target. These measures were substantial and far-reaching. They included major commitments of marketing activity in Japan by U.S. industry.[350] Collaborative efforts were undertaken to understand customer requirements and product capabilities. These were accompanied by competitions at the design-in level. Extensive follow-up and product support were also made available.

Supportive Policy Developments

A parallel development during this period very much affected the viability of the U.S. semiconductor industry: the emphasis on intellectual property protection (including the enactment of a chip protection law in the United States at the end of 1984 and in Japan beginning in 1986). Intellectual property protection is essential to commercial development in semiconductors and to the large R&D investments necessary to generate new products. The chip protection legislation created a new form of intellectual property protection for semiconductor products that was readily adopted internationally. It effectively reaffirmed the importance of general principles of trademark law and reduced the risks of piracy in semiconductor design. In addition, despite considerable controversy within the U.S. government, the decision to establish SEMATECH, a consortium of U.S. semiconductor manufacturers, backed by the Department of Defense and a substantial annual commitment of funds, sent a powerful signal to foreign producers of semiconductors that the U.S. government was not prepared to see the U.S. semiconductor industry marginalized. Both these developments increased the confidence of the U.S. industry and the trade policy community that the cost-based, market access approach of the STA would have an opportunity to work.

Success of the Trade Agreement

The STA was an unprecedented mechanism to increase competition and reduce friction in a key high-technology industry. Without this effective trade defense, combined with the STA's innovative market-opening mea-

---

[350] The Council on Competitiveness study emphasized the role of marketing efforts of U.S. semiconductor producers, noting that "as a signal of their commitment to the Japanese market, 125 U.S. semiconductor companies established a presence in the country, and over the 1986–1992 period, U.S. firms opened an average of one new facility per month in Japan. In general, U.S. industry undertook a wide variety of efforts to market its products in Japan and made great efforts to improve its knowledge of, and exposure to, the Japanese market." Council on Competitiveness, *Roadmap for Results,* p. 54.

sures, the structure and character of the U.S. industry would be very different today, and its contribution to the U.S. economy significantly smaller.[351] Indeed, some informed participants believe both that the U.S. industry should have forged a collective response earlier and that the government, in turn, should have been more responsive to the needs of this strategic industry. While the policy process ran its course, the U.S. industry suffered significant damage; thousands of jobs and billions of dollars in revenue were lost before the U.S. responded.[352]

Although some of the elements of the STA remain controversial, overall the agreement has been very positive.[353] Costly and destructive dumping has ceased. U.S. producers of DRAMs and EPROMs have survived and rebounded as world-class competitors. Major new entrants into DRAM production, notably producers in Korea and, increasingly, Taiwan, have substantially increased competitive vitality in the world DRAM market. U.S. and foreign producers have achieved far greater access to the Japanese market. While this progress remains vulnerable to arbitrary reversal, its underpinnings are market based. It has provided both major long-term benefits to global consumers and a competitive environment in which the U.S. industry could recover and prosper.

An Agreement Oriented to High-Technology Competition

In part, the success of the STA can be attributed to its sensitivity to the peculiar features of high-technology competition. Competition had to be preserved rather than merely offset by government intervention. By requiring future competition to be cost based, the STA strengthened market competition while minimizing government intervention. However, swift and

---

[351] Ibid., p. 42.

[352] Laura Tyson, *Who's Bashing Whom?* p. 271, and the Council on Competitiveness study emphasize the special characteristics of high-technology industries and the need for quick decisionmaking processes in responding to trade problems. The report adds that "the U.S. semiconductor industry may have been lucky to survive delay. Other technology-intensive industries may not be." *Roadmap for Results,* p. 55.

[353] Some observers argue that the agreement was a failure, resulting in significantly higher prices for U.S. consumers of semiconductors and in a windfall gain in profits for Japanese producers. In this view, the agreement's "bubble profits" for the Japanese producers were plowed back into R&D investment, having the "perverse result of strengthening the Japanese companies for future rounds of competition in new products." (Laura Tyson, *Who's Bashing Whom?* p. 117.) For a critique of the agreement, see David Mowery and Nathan Rosenberg, "New Developments in U.S. Technology Policy: Implications of Competitiveness and International Trade Policy." *California Management Review,* vol. 32, no. 1, Fall 1989, pp. 107–124. For a positive assessment, see Thomas Howell, et al, *Creating Advantage,* Laura Tyson also rejects many of these criticisms. See Laura Tyson, *Who's Bashing Whom?* p. 87 and p. 132 cited below.

predictable government action was a necessary condition of success. The STA provided it in the form of a mechanism that was capable of detecting and combating dumping before the targeted firms were effectively removed as players in global competition. The willingness of the U.S. government to take action, as noted above, was a crucial condition for success.[354]

Moreover, the policy approach adopted by the American government had the considerable advantage of being genuinely international. The STA accomplished this by taking a comprehensive approach that addressed, for example, third-country dumping and sought market access for all non-Japanese firms, not just U.S. firms. Finally, the arrangement had to be based on market realities. The needs of U.S. and Japanese customers were taken into account, and genuine industry-to-industry interaction, design-ins and technology adaptation were undertaken.

This progress was a joint effort. In part, it was facilitated by the expanded commitment of U.S. semiconductor firms in Japan. The number of technical personnel was significantly increased, and new design centers were opened at the time of the conclusion of the second Semiconductor Agreement. However, MITI played a determining role, urging Japanese semiconductor producers to increase procurement of foreign components and to submit reports with purchasing plans for foreign semiconductors as a means of expanding foreign market share.[355]

The phenomenal growth of the world market for semiconductors has probably contributed to the success of the agreement, not least by reducing the zero-sum nature of its results. This has been especially true in recent years. For example, the world market for semiconductors in 1993 was on the order of $77 billion. By 1995, the market had nearly doubled, to $146 billion.[356] Aided by this high-growth environment, progress under the agreement in terms of foreign share of the Japanese market was significant, rising in 1995

---

[354] Thomas Howell argues that the delay in fulfilling the market access commitment made in 1986, until the U.S. retaliated in April 1987, should not be interpreted as proof that the Japanese respond only when confronted with sanctions. He suggests that the explanation is more complicated, noting that Americans tend to see the negotiated solution as final, with implementation flowing naturally, while Japanese tend to see the negotiated solution as one more stage in the negotiations, with implementation a subject of further negotiations. Howell et al., *Creating Advantage,* p. 89. It might be added that, given the lack of continuity in the upper levels of the U.S. government, the implementation negotiation is likely to be with a new American negotiating team, possibly having no prior experience with the subject at hand.

[355] Howell et al., *Creating Advantage,* p. 93, citing Japanese press reports. See also p. 84.

[356] Semiconductor Industry Association, *World Semiconductor Forecast,* San Jose, Calif., November 1995. Interestingly, U.S. market share has declined from a peak of 43 percent to 41 percent in 1995. Japanese market share is estimated at 39 percent, with the substantial remainder made up by the rest of the world, notably Korean, Taiwanese, and European production of DRAMs. The DRAM market showed a phenomenal increase, rising from $13 billion in 1993 to $41 billion by 1995. Ibid.

to a fourth quarter peak of 30 percent, though some of this increase results from shifts in product mix toward microprocessors, with DRAM-intensive Japanese consumer electronics production moving off-shore. The benefits of this increased market share for U.S. producers were significant.[357]

Some U.S. and foreign producers, and their customers, were disadvantaged by price increases and especially by delay in product delivery.[358] It is important to recognize, however, that these trends were under way *before* the trade agreement took effect, and reflected the characteristic "market counter-measures" taken by many Japanese industries in depressed markets.[359] The shortage in supply of DRAMs did prove highly profitable for Japanese semiconductor producers. From a long-term perspective, however, U.S. and foreign producers, their equipment suppliers, their employees, and downstream users of semiconductors were advantaged, while global competition in this enabling technology was strengthened.[360]

Some critics of the Semiconductor Agreement were less concerned with the agreement's market effects than with a question of principle, on the

---

[357] While the gains for the U.S. economy from the agreement are difficult to calculate with precision, they are significant. The increased market share translates into R&D and capital investment expenditures as well as high-wage employment. At the end of 1990 the SIA calculated that U.S. firms achieved an incremental $1.16 billion in annual revenues under the STA (and would presumably have achieved an additional $1.16 billion had the 20 percent target been reached). A revenue gain of this magnitude ($1.16 billion) permits increased investment of an estimated $137 million annually in R&D, and of $130 million in annual capital expenditures, which in turn stimulates the sales of U.S.-based semiconductor equipment and materials suppliers. Howell et al., *Creating Advantage,* p. 89.

[358] Howell et. al., *Creating Advantage,* p. 130. Howell documents cases where European producers reported that Japanese suppliers withheld critical components and machines for competitive purposes during this period.

[359] Laura Tyson, *Who's Bashing Whom?* p. 117. Howell et al., *Creating Advantage,* pp. 116–132. Howell argues that "the evidentiary record shows that joint actions by Japanese firms were under way in 1985, well before the [Semiconductor] Arrangement existed; the precipitating event, to the extent one may have been needed to prompt joint market-regulating actions, appears to have been the elimination of most non-Japanese producers of DRAMs in mid-1985." Ibid., p. 129. Recent analysis supports this view. Kenneth Flamm notes that "Japanese semiconductor analysts in Tokyo stressed both the trend toward more oligopolistic behavior in the chip industry and the [Japanese] government's role in encouraging it." See Kenneth Flamm, *Mismanaged Trade,* p. 206. Flamm also argues that the production cuts and restraint in investment associated with MITI's implementation of the STA prior to 1988 subsequently "played some role" in the increase in chip prices by reducing aggregate world supply. He adds, however, that "the brakes applied to Japanese supply apparently greatly exceeded the restraints required" to meet the requirements of the Commerce Department. Ibid., pp. 269–272.

[360] Laura Tyson supports this assessment. She argues that "contrary to popular belief, the agreement was not ineffective at realizing many of its aims, including those of stabilizing the share of U.S. producers in the global DRAM market, reversing the precipitous decline of the U.S. share in the global market for EPROMs and increasing the share of U.S. producers in the

grounds that governments should not directly influence purchasing decisions of private enterprises. The 20 percent market share target was singled out as an example of "managed trade." This criticism, though widespread, presumed that the rest of Japan's import trade was unrestrained. Industry experts argued that Japan's high-technology imports were already managed, either directly by MITI or through informal government-industry arrangements with industrial groups and distributors, or through private anticompetitive practices.[361] In this view, what was unusual about the Semiconductor Trade Agreement was not that it included a market share target, but that its management was both public and jointly administered (i.e., by the Japanese *and* American governments).

Leaving aside ideological objections, the Semiconductor Agreement has the not inconsiderable virtue of having accomplished its goals. It offered an innovative, market-opening solution to a previously intractable competitive problem for the American industry. The negotiation of the STA and the commitment of the U.S. government—and ultimately the Japanese government—permitted the removal of a potentially debilitating source of friction from a key bilateral relationship. Moreover, by providing an effective means for overcoming the trade problems that threatened the viability of the U.S. industry and for restoring cost-based competition, the agreement established a durable basis for vigorous global competition and allowed the semiconductor industry to continue its phenomenal growth from an increasingly dispersed production base.

SUPPLEMENT B.
GOVERNMENT SUPPORT FOR TECHNOLOGY
DEVELOPMENT: THE SEMATECH EXPERIMENT[362]

Overview

The U.S. response to the challenge to its semiconductor industry in the period 1985–1995 involved three inter-related elements. As noted in the

---

Japanese market. Nor did the agreement reduce competition across the board, as is widely believed. Instead the effects on competition also varied by industry segment over time." *Who's Bashing Whom?* p. 87 and p. 132.

[361] Howell et al., *Creating Advantage,* pp. 101–103. MITI's subsequent success in setting minimum pricing standards for Japanese DRAM manufacturers' direct export sales lends support to this view. See Kenneth Flamm, *Mismanaged Trade?* p. 243.

[362] This section draws heavily from the contributions to Committee deliberations of George M. Scalise, *The Nature of High Technology Competition,* and William J. Spencer, *SEMATECH, passim.* See also William J. Spencer and Peter Grindley, "SEMATECH after Five Years: High-Technology Consortia and U.S. Competitiveness," *California Management Review, vol. 35,* Summer 1993.

section on the Semiconductor Trade Agreement, the first critical assistance took the form of trade relief. A second element was the creation of the consortium of U.S. chip producers and equipment manufacturers that came to be known as SEMATECH. However, the third indispensable element in the revival of the U.S. industry rested on its willingness and ability to reinvest and innovate.

Arguments about which element was most decisive probably miss the point. The recovery of the American industry was achieved by a combination of private effort and public-private interactions which enabled the U.S. firms to profit from shifts in demand, i.e., away from DRAMs (where Japanese skill in precision clean manufacturing gave significant advantage) toward microprocessor design and production (where U.S. strengths in software systems and logic design aided their recovery).[363] The recovery of the U.S. industry is thus like a three-legged stool. It is unlikely any one factor would have proved sufficient independently.[364] Trade policy, no matter how innovative, could not have met the requirement to improve U.S. product quality. On the other hand, by their long-term nature, even effective industry-government partnerships can be rendered useless in a market unprotected against predatory pricing by foreign rivals. Most importantly, neither trade nor technology policy can succeed in the absence of adaptable, adequately capitalized, effectively managed, technologically innovative companies. Because the rest of this section addresses "one leg of the stool," i.e., the SEMATECH contribution, it is essential to underscore the importance of the other two legs, which are an effective trade policy and a dynamic private-sector able to attract and retain the capital necessary to compete in global markets.

The Technology Challenge

In addition to the challenge posed by Japanese trade practices (see Supplement A), the U.S. industry faced major technology problems. In the early 1980s, U.S. manufacturing quality had fallen behind the standards of the Japanese industry. The problem was manifest in lower manufacturing yields, higher costs, and inferior product quality.

---

[363] Charles H. Fine and Daniel E. Whitney, "Is the Make-Buy Decision Process a Core Competence?"

[364] Michael Porter captures the point, noting that "(g)overnment cannot create competitive industries, firms must do so. Government's role in competition is inherently partial, because many other characteristics of a nation bear on it. Government can shape or influence the *context and institutional structure* surrounding firms, however, as well as the *inputs* they draw upon. Government policies that succeed are those that create an environment in which firms can gain competitive advantage..." Michael Porter, *The Competitive Advantage of Nations,* p. 620.

The Japanese industry had established great strength in manufacturing, whereas the U.S. industry excelled in product design. As the Japanese industry expanded during the early 1980s, particularly in the product categories that involved high-volume manufacturing, their advantage in manufacturing grew steadily. While the manufacturing technology problem for U.S. semiconductor producers was greatly aggravated by the consequences of the Japanese trade practices, it nonetheless constituted a distinctly different challenge, requiring its own solution if the U.S. industry was to exploit the opportunities provided by the Semiconductor Trade Agreement.

Because the quality technology problem was centered on process technology and the performance of the supporting infrastructure of tools and materials, the manufacturing challenge was common to all U.S. device makers. It was also once removed from direct product or market competition in semiconductors, because it involved technology that was not part of device design or specific device production recipes.

The scope of this manufacturing technology challenge was too great for any single company to overcome on its own. As a result, individual companies gradually came to realize that only a consortium that garnered the participation of all those who would benefit from solving the manufacturing challenge was likely to be able to afford the undertaking and make it succeed.[365]   In 1987 fourteen U.S. semiconductor manufacturing firms, accounting for over three-quarters of U.S. production, came together to form a consortium to develop semiconductor manufacturing tools, materials, and technology.[366]

U.S. Government Interest

By the mid-1980s, U.S. officials had become keenly aware of the problems besetting the semiconductor industry and of the implications of the industry's decline both for economic growth and national security. There was growing concern among senior industrialists and defense officials over the erosion of the U.S. ability to produce key components of critical weap-

---

[365] The Semiconductor Industry Association put forward a proposal in May 1987 for a research consortium supported by shared government-industry funding.

[366] Eleven of the original fourteen are still members. Dissatisfaction with the focus of the research program led Micron Technology and LSI Logic to withdraw in 1992. In 1993, Harris Corporation withdrew. For a discussion of the SEMATECH R&D program, see the GAO study *SEMATECH's Technological Progress and Proposed R&D Program,* GAO/RCEED/92-223 BR, Washington, D.C., July 1992. Current SEMATECH membership includes Advanced Micro Devices, Digital Equipment Corporation, Hewlett-Packard Company, Intel Corporation, IBM, Lucent Technologies, Motorola, National Semiconductor, NCR, Rockwell International, and Texas Instruments.

ons systems. The national security implications of the growth in U.S. dependency became apparent when Norm Augustine, chairman of the Defense Science Board and of the Martin Marietta Corporation, informed Secretary of Defense Caspar Weinberger that twenty-one absolutely critical U.S. military systems contained chips available only from foreign producers.[367] Moreover, defense industry leaders such as Augustine believed that U.S. military suppliers should be able to rely on cost-effective U.S. sources for the manufacture of advanced designs or high-volume, low-cost reliable commercial products.

Importantly, the Japanese producers were displacing not only the U.S. semiconductor manufacturers, but also the equipment and materials manufacturers who supplied them. Because advances in semiconductor technology are closely linked to equipment and manufacturing capabilities, this meant the United States was losing its capacity in the latest manufacturing technologies—critical determinants of price and quality. This erosion in U.S. technological competency also meant that the United States was well on its way to becoming structurally dependent on Japan for technology critical to the U.S. defense capability.[368]

In light of these developments, the Defense Department concluded the U.S. industrial base needed the manufacturing technologies that the consortium proposed to address. The proposal was closely aligned with DOD's own plans for the development of semiconductor manufacturing technology, especially as it was completing a five-year program in the design of integrated circuits for military applications. This technology could allow the Defense Department to deploy advanced integrated circuits in virtually all of its major defense systems in a more effective and cost-efficient manner. By participating with the consortium rather than seeking to procure R&D through normal defense contracting procedures, the Defense Department believed it could achieve the required capability while substantially reducing its costs.

To capture these savings, benefit from the anticipated improvements in semiconductor manufacturing technology, and bolster the defense industrial base, the Defense Department supported the consortium with approximately $100 million a year, on the condition that industry match the funding with

---

[367] Clyde Prestowitz, *Trading Places,* p. 122.

[368] Ibid. The National Advisory Committee on Semiconductors confirmed this assessment, stating that "if this vital industry is allowed to wither away, the nation will pay a price measured in millions of jobs across the entire electronics field, technological leadership in many allied industries such as telecommunications and computers, and the technical edge we depend on for national security." See *A Strategic Industry at Risk,* November 1989, cited in Council on Competitiveness, *Roadmap for Results,* p. 39.

its own contributions.[369] On this matched-funding basis with industry, Defense Department financial support continued for eight years. In July 1994, the SEMATECH Board voted to terminate government support for its effort, with fiscal year 1996 slated to be the final year of government funding. Despite the end to government funding, SEMATECH members have agreed to continue this innovative program with core funding provided exclusively by corporate contributions.

The SEMATECH Experiment

SEMATECH represents a significant new experiment for government-industry cooperation in technology development.[370] Conceived and funded under the auspices of the Reagan administration, the consortium represented an unusual collaboration for both the U.S. government and the U.S. semiconductor industry. In many respects, the decision to cooperate represented a profound change for the highly competitive, market-oriented companies that formed its membership. Indeed, the move to form SEMATECH occurred only when the top executives from the best U.S. electronics companies came to the sobering realization that they could not prevail alone. As one commentator noted, "the mere formation of SEMATECH required a radically new mind-set at some of America's leading high-tech corporations."[371] Not only were these independent executives now willing to collaborate with the government, but these fiercely competitive rivals had to agree to collaborate on research and development that was competitively important to each of them.

A Short-Term Focus on Improved Quality

The technology focus of the consortium is near-term and of almost immediate commercial significance. It seeks to affect existing manufacturing processes, tools, and materials. However, no member company or member company product receives any financial assistance. For member companies, the consortium provides generic manufacturing technology and improved tools that allow them to compete more effectively. For equipment and materials suppliers, it provides competitive awards, customer input, and industry-wide demonstration and validation of products. Overall, these measures have had a positive, competitive effect on the U.S. supply base.

---

[369] Membership dues have been fixed at one percent of semiconductor sales, with a minimum contribution of $1 million and a maximum of $15 million.

[370] For an excellent review of this experiment, see Peter Grindley et al., "SEMATECH and Collaborative Research: Lessons in the Design of High-Technology Consortia."

[371] Hedrick Smith, *Rethinking America.* Random House, New York, 1995, p. 385.

Industry Driven

From the outset, leadership for the consortium was provided by industry. The CEO of National Semiconductor, Charles Sporck, provided crucial impetus to the concept of the consortium, playing an instrumental role in garnering support among the original fourteen members of SEMATECH. The consortium also benefited from the prestige and leadership of Bob Noyce, who strengthened the credibility of the endeavor within the industry and in Washington.

Reflecting these origins, the operations of SEMATECH have remained industry driven. While it continues to benefit from strong leadership, no single entity dominates the consortium or determines its direction. Members, including Department of Defense officials, reach consensus on technical direction; the consortium management is then left to implement that direction. The industry interaction within the consortium changed the dynamics between device makers and suppliers, with the collaboration generating new technical perspectives for the participants and encouraging the give-and-take between manufacturer and supplier so as to enrich, improve, and expedite the technology development process.

Because SEMATECH programs are focused on SEmiconductor MAnufacturing TECHnology (hence SEMATECH's name and its mission), the largest portion of its budget is spent on programs which relate to equipment improvement. Quite often these programs are carried out at equipment supplier sites or at member company sites. Teams of SEMATECH engineers are involved in these programs, as well as the on-site engineers. Technology diffusion is thus achieved through first-hand transfer. Member companies provide technical personnel to work at the consortium and regularly rotate them back to the member company.[372] By having a member's output from SEMATECH depend largely on the quality of the people assigned to the consortium, technology transfer is enhanced and there is a strong incentive to provide high-quality people to the consortium.[373]

Closed Membership—Open Output

Membership in SEMATECH and in the group of participating supplier firms known as SEMI-SEMATECH is limited to U.S.-based companies. This

---

[372] "In 1992, some 225 of the 722 personnel employed at SEMATECH were assignees from member firms. Most of these individuals were members of the consortium's 300 member professional research staff." See Grindley et al., "SEMATECH and Collaborative Research," p. 729.

[373] SEMATECH also produces reports and holds meetings to provide technology information to member companies. For example, in 1994 SEMATECH held over 600 meetings and hosted over 25,000 visitors. See Spencer, *Technology (Transfer) at SEMATECH,* 14 December 1995 contribution to Steering Committee deliberations, p. 1.

preference was based on the corporate restrictions imposed by the U.S. industry participants, although the policy had the tacit support of U.S. Defense Department officials. This limitation was also probably necessary in order to maintain political support for continued government funding.[374] Originally, the results of SEMATECH research were licensed to member firms only on an exclusive basis, although that requirement was changed in 1991. At present, the output of the consortium is *not* restricted to U.S. entities, although members of the consortium do receive some priority in ordering and receiving equipment derived from innovations funded by SEMATECH.[375] Currently, members and supplier firms alike are encouraged to deploy SEMATECH technology throughout the world. Moreover, new projects, like the I300I initiative to develop an industry standard for the move to the 300mm wafer, involve extensive international participation.[376]

The Semiconductor Roadmap

The scope and success of this initiative ultimately led to a systematic review of manufacturing needs across the entire technology spectrum. The Semiconductor Roadmap exercise brought together, in a systematic fashion, expert representatives of industry, universities, and government to examine the technological challenges facing the industry. The Semiconductor Roadmap represents an unprecedented collective effort to identify opportunities and obstacles to the continued exponential growth in performance of the industry's products. SEMATECH's efforts became the basis for the development of

---

[374] Andrew Procassini, the former president of the Semiconductor Industry Association, confirms this point in *Competitors in Alliance,* p. 74. See also Grindley et al., "SEMATECH and Collaborative Research," p. 747. The authors note that SEMATECH's May 1993 announcement that it would continue to work with a U.S. equipment manufacturer despite an impending alliance with a major foreign competitor was criticized by the House Armed Services Committee. The Committee questioned "the rationale of Congressionally initiated investments if these investments do not result in jobs and industry in the U.S." See also J. Robertson and J. Doresch, "GCA Revives as U.S. Drafts Corning, AT&T." *Electronics News,* 16 August 1993 cited in Grindley et al., "SEMATECH and Collaborative Research."

[375] Grindley et al., "SEMATECH and Collaborative Research," p. 731. The authors note that the change in disclosure policies both benefited U.S. equipment manufacturers and addressed criticism from some semiconductor manufacturers about the use of general tax revenues to benefit member companies at the expense of nonmembers. Ibid. This policy has also enhanced the ability of nonmembers to obtain a "free-ride," that is benefit from SEMATECH advances without contributing to the program. P. 745.

[376] As noted above, Japanese producers did not join the international effort organized by SEMATECH to develop the 300mm wafer, preferring a separate national effort, though some information will be exchanged between the two projects. Kenneth Flamm, *Mismanaged Trade?* p. 441. The author also suggests that SEMATECH's sponsorship of the I300I program is a move toward greater international cooperation made possible by the absence of the constraints associated with government funding.

aggregate technology roadmaps that could identify technical opportunities and gaps that had not previously been brought to light.

Was SEMATECH A Success?

An assessment of an R&D program such as SEMATECH is a difficult analytical undertaking, not least because of the "counter-factual challenge" of knowing what would have happened to the industry without SEMATECH—keeping in mind that the consortium's technological accomplishments were widely disseminated to both participants and nonparticipants. Moreover, the success of R&D efforts are difficult to measure with precision.[377] The criteria are elusive, the output can rarely be quantified, and know-how or technology is only one—and not necessarily the most decisive—element in the product development and commercialization process.[378] Despite these inherent uncertainties, the assessments summarized below suggest a record of substantial accomplishment.[379]

Industry Goals Met

From the industry perspective, SEMATECH's goals have been largely achieved.[380] The manufacturing quality and efficiency of the U.S. industry have at least reestablished parity with Japan. With a few notable exceptions, at least one world-class U.S. supplier exists in all the primary segments of the semiconductor manufacturing process.[381] The worldwide mar-

---

[377] Assessing technological success, and a consortium's contribution to technological advance, can be complex. Success has many fathers. Moreover, the contribution of technology to commercial success is difficult to assess, especially for government-funded consortia. As Grindley, Mowery, and Silverman note "the politically relevant timetable for such evaluation is often far shorter than the period of time needed for the realization of the economic effects of consortia research." Grindley et al., "SEMATECH and Collaborative Research," p. 724.

[378] Ibid. In reviewing the difficulties SEMATECH encountered in maintaining a viable U.S. producer of lithographic equipment, the authors underscore the argument that "technology alone is rarely sufficient to restore the competitiveness of firms or industries that lack critical, complementary personnel, marketing, managerial, and financial resources." P. 743 and p. 751.

[379] Ibid., p. 736 and passim. The authors discuss the challenges of evaluating SEMATECH and provide a critical, but generally positive, assessment of the consortium's accomplishments. They also note that the consortium's goals changed over time, reflecting the changing perceptions of its members' needs. This operational flexibility is a strength and probably essential in an industry evolving as rapidly as the semiconductor industry.

[380] For example, SEMATECH has met its objectives in the development of process technology, the supply of manufacturing equipment, and collaboration between manufacturers, suppliers, and research centers. Ibid., p. 743.

[381] Relative to both Europe and Japan, the United States lags in terms of company rankings in sectors such as packaging equipment, mircrolithography equipment, and semiconductor materials. Presentation by VLSI Research Inc., President's Conference, 11 April 1995, Washington, D.C.

ket share decline of both U.S. semiconductor firms and suppliers has been reversed and worldwide market share leadership regained. While other factors were also in play, many in the industry believe SEMATECH made a major contribution to these turnarounds.

The consortium certainly achieved its initial objective to match and then overtake its competitors in semiconductor manufacturing technology. U.S. manufacturers were able to construct integrated circuits with features as small as 0.8 micron on five-inch silicon wafers by 1989, 0.5 micron features on a six-inch wafer in 1990, and 0.35 micron features by the end of 1992. Keeping in mind that a 0.35 micron particle is approximately 1/200th the width of a human hair, this is a remarkable technical achievement. As noted in the section on international cooperation above, SEMATECH is now playing an instrumental role in the industry move to the 300mm wafer standard.

The consortium itself has taken the view that an important part of the programs is the measurement of results. Earlier in its history, SEMATECH established a return on investment measurement for each program. The average return tended to be between 3.5 and 4 times for the member companies, with a range of return from about 8 to about 2 times investment.[382] In 1994 a new measurement system was established measuring user satisfaction and user support for each program. Plotted in four quadrants with the quadrants divided at the 50 percent line for customer support and customer satisfaction, the goals for SEMATECH programs are to have 70 percent in the third quadrant, where customer support and customer satisfaction are both over 50 percent. Those programs that fall in the first quadrant are candidates for termination.

The creation of SEMATECH encouraged technology-based competition among equipment and materials suppliers. Rather than alter market forces, SEMATECH built on them. Consequently, government-industry collaboration on semiconductor technologies did not fundamentally undermine market forces. Inefficient companies were not propped up by government funds, and assured commercial success was not bestowed on any consortium member. Instead, this innovative collaboration offered a source of technology that would then have to be effectively commercialized before a firm could extract any economic benefit for itself.[383] SEMATECH's role in supporting equipment development and qualification does mean that it has substantial influence on technological choices in an industry faced by great uncer-

---

[382] Spencer, *Technology (Transfer) at SEMATECH*, p. 1.

[383] The impact of government support for technology is generally believed to be minimally distortive of the market because it tends to occur at the beginning of the product development cycle rather than at the end. This is the rationale for the current exemptions in the GATT for government R&D support, as noted in the body of the report.

tainty. However, SEMATECH's private sector management has sought to maintain a diversified portfolio of technological alternatives and has hedged against technological risk by pursuing alternative technologies with more than one supplier and more than one engineering approach.[384]

Perhaps the best measure of SEMATECH's success for the industry is the continued willingness of members to invest in the consortium even without government participation. Moreover, the improved technology and more capable suppliers that became available through SEMATECH's efforts supported not only U.S. semiconductor manufacturers but the emerging industries in Korea and Taiwan as well. European and Japanese industries were also able to draw on the results of consortium research. The technical accomplishments of SEMATECH resulted in higher productivity, better product, and a strengthened U.S. supply base while contributing to the diversity of equipment supply which is essential for the maintenance of vigorous global competition.

DOD Goals Met

The Defense Department's return from SEMATECH can be similarly assessed. The Defense Department benefited from cost and quality improvements in the microelectronics that increasingly account for the performance advantage of U.S. defense systems. In addition, the vitality of the supporting U.S. infrastructure was preserved, and concentrated dependency on foreign sources of semiconductor tools and materials was avoided. Like other defense R&D programs, SEMATECH delivered technology that enabled the Defense Department to do its mission better.

University Spillovers

Even though government participation in the consortium was justified and funded as a defense program, broader national interests also benefited. SEMATECH provided a focus and exchange of technical ideas that, as it did for industry, strengthened ongoing R&D in both universities and government agencies and laboratories. For example, SEMATECH helped create, in cooperation with the Semiconductor Research Corporation, the university centers of excellence for semiconductor research. In addition, the Semiconductor Roadmap exercise mentioned above contributed a great deal to the identification of key research areas.

---

[384] Grindley et al., "SEMATECH and Collaborative Research," pp. 745–745.

National Economic Benefits

Perhaps most important, SEMATECH helped to anchor the future development of a strategic technology in the United States. This makes it likely that the United States will be the hub for subsequent deployment of the technology throughout the world. The United States is thereby assured of capturing a significant proportion of the economic growth, high-wage employment, and technological competency to be gained from the phenomenal growth that characterizes this industry. More broadly, the United States will continue to maintain and benefit from the economic leadership, independence, and prosperity that are uniquely associated with the mastery of high-technology manufacturing.

BOX J. LESSONS IN TECHNOLOGY TRANSFER

SEMATECH management has identified several principles of operations which underpin successful technology transfer activities. They include:

1. **User Involvement:** It is important to have the users involved in the determination of the programs, and the continual evaluation of the programs. These users have the responsibility for returning that technology to their member companies.

2. **Metrics:** Specific metrics for each program are essential for successful technology development. For example, programs are evaluated quarterly at SEMATECH.

3. **Direct Involvement of Senior Management:** This is essential to stop what does not work. Engineers quite often believe that if a program can continue for only a short time longer, the problems can be solved and the program will turn out to be of great value. Usually senior management in the member companies of SEMATECH is more inclined to stop programs. Consequently, the involvement of senior management is essential to close out programs with inadequate returns.

4. **High Return:** Focus on those programs where there is a high return. The users of technology are the best judges of rates of return. Support and satisfaction are one measure; SEMATECH has also experimented with return on investment. There are certainly other metrics that can be used. What is essential is that there be an accepted metric to determine which programs should be continued and which should be stopped.

5. **Effective Selection:** Over a period of nearly eight years, SEMATECH has developed a series of processes for choosing technology development programs which are relevant to the core of interests of the member companies. These processes are continually evaluated and updated. The consortium's goal is to find the best processes from anywhere in the world and apply them to its programs.

SUPPLEMENT C.
IMPLICATIONS OF THE U.S. DUAL-USE STRATEGY

The rapid advance of commercial technologies has led U.S. officials to push for a new defense acquisition strategy which emphasizes the use of new technologies, developed by commercial industry, to meet defense needs. While announced with some fanfare in the United States, in the view of some observers dual-use policies have in fact been practiced by others in the Cold War period, perhaps most effectively by Japan, and is now the policy of many European countries.[385] It is U.S. policy which, in the post–Cold War world, is evolving toward the practice of the other industrial powers.

The dual-use strategy adopted by the United States early in the Clinton administration has two key elements. First, the Department of Defense is to make use, wherever possible, of components, technologies, and subsystems developed by commercial industry. The Department is to develop defense-unique products only where necessary.[386] Second, for this acquisition strategy to be successful, defense R&D efforts must nurture technologies and capabilities that will continue to be advanced through industries' efforts to remain competitive in commercial markets. The goal of the strategy is to leverage commercial technological advance in order to create military advantage, while ensuring the resulting equipment remains both affordable and the most advanced in the world.[387]

In the words of one senior defense official, the objective of this policy is "to marry the momentum of a rigorous, productive, and competitive commercial industrial infrastructure with the unique technologies and system integration capabilities provided by our defense-industrial base."[388]   This policy represents a sharp departure from the deeply ingrained practice of defense funded R&D, with procurement and production consigned to specialized defense suppliers. In the past, when defense requirements constituted a significant portion of high-technology markets, this policy was often effective. Indeed, government-military procurement was instrumental in accelerating innovation in aerospace and even in generating new industries

---

[385] Presentation by Jacques Gansler to the conference *Sources of International Friction and Cooperation in High-Technology Development and Trade, 30–31 May 1995.* For a discussion of the need for the U.S. military-industrial complex to adjust to the end of the Cold War and maintain technological superiority with a reduced defense budget that nonetheless enhances rather than reduces U.S. economic competitiveness, see Jacques Gansler, *Defense Conversion Transforming the Arsenal of Democracy*, MIT Press, Cambridge, Mass., 1995.

[386] Flat Panel Display Task Force, *Building U.S. Capabilities in Flat Panel Displays, Final Report*, U.S. Department of Defense, Washington, D.C., October 1994, p. 1.1.

[387] Presentation by Under Secretary of Defense Paul Kaminski, in *Sources of International Friction and Cooperation in High-Technology Development and Trade, 30–31 May 1995* and the presentation by Stephen A. Cooney to the conference *Towards A New Global Framework for High-Technology Competition, 30–31 August 1995,* Kiel, Germany.

[388] Kaminski, op. cit.

such as microelectronics. However, the military emphasis on "the extremes of performance," with mission capability given first priority and cost being a secondary consideration, led military manufacturers to emphasize high-cost custom production rather than the low-cost, high-volume production that characterizes commercial markets.[389]

Some suggest this led military contractors "to lose the capability to design, manufacture, and sell in a competitive marketplace."[390] This less competitive, performance-only orientation had been identified by the 1986 Packard Commission, which stated that commercial industry could provide cost savings and quality improvements for military programs through the application of civilian technology.[391] Because the commercial sector relies on advanced technology both to lower costs and to enhance performance, through their interaction, cost *and* performance have improved faster in the commercial sector than for military applications. Moreover, the high development costs for new technologies are best supported by high-volume sales of consumer products. In addition, the acceleration of product life-cycles in competitive markets means that commercial producers have become significantly more responsive and flexible than military producers.[392] As a result of these trends, the Defense Department no longer leads in critical technologies such as information systems, telecommunications systems, microelectronics, and a variety of fields associated with advanced design tools.[393] Senior defense officials now assert that declining defense budgets, demand levels, and resources make reliance on defense-only suppliers both unaffordable and ineffective.[394]

To accommodate these trends, current U.S. policy is to rely on commercial producers for leading-edge capabilities, at reduced cost, in a shorter timeframe. Insertion time is increasingly important, because in the global markets which characterize many new technologies, access to the same commercial technology base by both allies and adversaries will become the norm. In this environment, military advantage will accrue to the nation with the best "cycle time," that is, the ability to acquire leading-edge technologies at low cost, insert them in weapons systems, and field soldiers trained to exploit them effectively.[395] If successful, this approach will

---

[389] Richard Samuels, *Rich Nation, Strong Army,* pp. 22–24.

[390] Ibid. Samuels also cites Seymour Melman (ed.), *The War Economy in the United States: Readings on Military Industry and Economy,* Saint Martin's Press, New York, 1971, p. 72.

[391] Richard Samuels, *Rich Nation, Strong Army,* p. 27.

[392] Ibid., p. 28.

[393] Paul Kaminski in *Sources of International Friction and Cooperation in High-Technology Development and Trade.*

[394] Flat Panel Display Task Force, *Building U.S. Capabilities in Flat Panel Displays.*

[395] Paul Kaminski in *Sources of International Friction and Cooperation in High-Technology Development and Trade.*

mean a sharp departure from the current fifteen-year acquisition cycle for major U.S. defense systems, and enable U.S. forces to benefit from the commercial turnover cycle of three to four years. For many systems, it will mean that the acquisition of defense-only technologies becomes the exception rather than the rule.

The adoption of a dual-use strategy by the U.S. also undercuts, at least in part, the widespread perception that U.S. defense programs and procurement served as "hidden subsidies" to the U.S. commercial sector. In part this view is widespread because U.S. defense and space agency officials sought to justify their budget requests with the argument that their programs had powerful and positive commercial benefits.[396] Further legitimacy was accorded this argument by countries directly subsidizing commercial technologies who saw U.S. defense budgets as a justification for their own commercially oriented efforts.[397] This claim was reinforced by companies responsible for defense work, which justify high defense budgets by asserting the benefits of defense acquisition. In some cases, there clearly have been benefits to U.S. industries such as aerospace and semiconductors—though the benefits of such support, especially in aerospace, seem to be somewhat unevenly distributed.[398] The validity of a direct comparison of the two types of support, the one commercially oriented from the outset (i.e., direct subsidies such as those for Airbus) and the other derivative (i.e., learning through military programs), is often overstated, particularly with respect to today's technologies and markets.

---

[396] Richard Samuels, *Rich Nation, Strong Army,* p. 21. See also Steven Rosen, *Winning the Next War: Innovation and the Modern Military,* Cornell University Press, Ithaca, N.Y., 1991.

[397] If the military procurement approach were best, presumably countries in Europe would have increased military budgets rather than direct support for commercial technology. Kende points out, for example, that "while many of the actions taken by the European governments in support of their computer industry were similar to those taken by the U.S. government, the goals were different, as were the results." Kende, "Government Support of the European Information Technology Industry," p. 4. In his analysis, Kende argues that the underlying concept of the national champion approach was flawed. For example, French, German, and British efforts in the 1960s to develop national champions to compete with IBM focused on company size per se, rather than focusing on the new technologies and quality products that characterized the American competition. He contends that the merger of several disparate national firms with incompatible product lines did not result in a dynamic global competitor—not least because governments removed the discipline of competition from the favored firms, which then benefited from protected national procurement markets.

[398] Some analysts argue that the unwillingness of the American government to intervene in the aerospace industry for commercial objectives has actually weakened U.S. producers, notably McDonnell Douglas. Laura Tyson, *Who's Bashing Whom?* p. 192. Tyson argues further that the scale, scope, and learning economies mean new entrants face much higher production costs than do incumbents. Without "the visible hand" of government support, the particular conditions which characterize the aircraft industry pose insurmountable barriers to new competitors. Ibid, p. 157.

Even in the heyday of defense led R&D, the assistance to U.S. industry derived from the U.S. defense budget was often constrained by the defense mission. Some analysts argue that the U.S. defense acquisition system, far from being a guise for the support of commercially relevant industry, has in fact created disincentives and barriers to the operation of market forces. These include "the unique government oversight requirements, the unique procurement requirements, (and) the unique military specifications" associated with military procurement.[399] Indeed, these restrictions are seen as so onerous that some world-class U.S. companies, e.g., Hewlett-Packard, refuse to accept defense R&D contracts.[400] Some analysts suggest that U.S. defense acquisition practices *lowered* overall return on U.S. R&D spending.[401] This is because the normal output of defense R&D "spun away" rather than "on" to commercial applications. Normal technology diffusion could not take place as a result of defense secrecy requirements.[402] Some analysts take the case a step further, not only outlining the shortcomings of the "spin-off paradigm," but arguing that massive defense spending did great harm to the [U.S.] economy.[403] This is because the defense industry has

---

[399] Presentation by Jacques Gansler in *Sources of International Friction and Cooperation in High-Technology Development and Trade.*

[400] Ibid.

[401] See Robert A. Solo, "Gearing Military R&D to Economic Growth," *Harvard Business Review,* November–December 1962, pp. 49–60, cited in Richard Samuels, *Rich Nation, Strong Army.* Samuels adds that "it is far from clear that the military acts primarily as an agent of technological innovation, that a linear relationship exists between defense and civilian products, or that military spending has had a positive impact on the civilian economy." Ibid., p. 22.

[402] "Governments have, of course, struggled to *prevent* diffusion of military technologies... This enforced isolation has taken a great many forms, including separate accounting and audit systems, different bidding and marketing procedures, export controls (extending to dual-use technologies), separate divisional structures, classification schemes, and nationality requirements, all of which aim to prevent (or at least to impede) technology diffusion." Ibid., p. 22.

[403] Exponents of this view cited by Samuels are Jay Stowsky, *Beating our Plowshares into Double-Edged Swords: The Impact of Pentagon Policies on the Commercialization of Advanced Technologies,* Berkeley Roundtable on International Economics, Berkeley, Calif., 1986; Robert Solo, "Gearing Military R&D to Economic Growth," *Harvard Business Review,* November–December 1962; Amitai Etzoini, *The Moon-Doggle,* Doubleday, New York, 1964; Seymour Melman et al., *The War Economy of the United States: Readings on Military Industry and Economy,* St. Martin's Press, New York, 1971; Mary Kaldor, *The Baroque Arsenal,* Hill and Wang, New York, 1981; James Fallows, *National Defense,* Random House, New York, 1981; Lloyd Dumas, *The Overburdened Economy,* University of California Press, Berkeley, Calif., 1986; Nathan Rosenberg, "Civilian 'Spillovers' from Military R&D Spending: The U.S. Experience Since World War II," in S. Lakoff and R. Willoughby (eds.), *Strategic Defense and the Western Alliance,* Lexington Books, Lexington, Mass., 1987; Jacques Gansler, "Integrating Civilian and Military Industry," *Issues in Science and Technology,* Fall 1988; John Alic et al. (eds.), *Beyond Spinoff,* Harvard Business School Press, Boston, Mass., 1992; and Jay Stowsky and Burgess Laird, "Conversion to Competitiveness: Making the Most of the National Labs," *American Prospect,* Fall 1992.

become progressively isolated from commercial industry, and because military and civilian laboratories compete (at least to some extent) for the same research personnel and research funding.[404]

Indeed, the evolution of U.S. policy toward a dual-use strategy essentially supports these claims. Instead of the commercial sector benefiting from technological spin-off, the military now requires commercial "spin-on." This strongly suggests that the previous U.S. approach, with the current barriers to the integration of the civil and military industrial bases, has proven less efficient than more commercially oriented strategies for a wide variety of defense components and equipment.

Consequences for Competition—

The more widespread adoption of dual-use policies is also likely to have important consequences for high-technology competition and cooperation. From the perspective of global competition, efforts by the United States and other nations to ensure an adequate commercial-industrial base to meet defense needs raise concerns that the defense rationale may be applied to a broad range of leading-edge industries, with defense requirements becoming an important source of financial support.[405] For example, government efforts to develop military weapons systems, such as the current Japanese production of the FSX, components for the Boeing 777, or the recently announced plan to produce an antisubmarine patrol plane, can have direct impact on commercially oriented aerospace products, much as U.S. Cold War military programs sometimes encouraged U.S. commercial aircraft development.[406] Because reliance on defense-unique industries is no longer cost effective, or even possible in some cases, the Defense Department appears prepared to foster, or at least encourage, a viable domestic industry that is globally competitive in terms of both cost and technological capability, as a means of meeting defense needs. The most striking example of this approach is the U.S. dual-use program on flat panel displays, an unusual effort to encourage a U.S.-based commercial industry—where none exists—to meet military needs from lower-cost, high-volume, commercially oriented production facilities.[407]

---

[404] Richard Samuels, *Rich Nation, Strong Army*, pp. 22–23.

[405] Sylvia Ostry, *"Technology Issues,"* paragraphs 21–24. Ostry cites in particular the evolution of U.S. policy toward dual-use technology. Richard Samuels, *Rich Nation, Strong Army,* argues that Japan has consciously pursued a dual-use policy for many years.

[406] As noted, aerospace spin-off can be overstated. See the section above, and the presentation by Sally Bath, director of aerospace, U.S. Department of Commerce in *Sources of International Friction and Cooperation in High-Technology Development and Trade.*

[407] For a description of this innovative program, see Flat Panel Display Task Force, *Building U.S. Capabilities in Flat Panel Displays.*

—and Cooperation

At the same time, the dual-use strategy includes an important cooperative element. In order to obtain affordable weapons systems, the U.S. defense strategy places increased emphasis on cooperation with U.S. allies in the acquisition of defense equipment. The renewed emphasis on cooperation rests on a combination of political, military, and economic rationales. Politically, the cooperative programs "help strengthen the connective tissues, the military-industrial relationships, that bind...nations in a strong security relationship."[408] The military rationale rests on the likelihood of combined operations in a coalition environment where performance can be considerably enhanced with interoperable equipment and rationalized logistics. And the economic rationale is simply that the budgets of the U.S. and allied militaries are shrinking. Consequently, what may not be affordable individually may be possible collectively.[409]

As with commercial alliances, international cooperation for military systems is not without its problems. These include the very long decision times often associated with cooperative projects, in which complex national decision processes have to be combined to produce a collective decision. This often results in "very poor cycle times and very long schedules."[410] Cooperative programs can also result in the absence of effective competition once an international team is assembled, which can lead to high cost and inefficient performance for weapons systems.

Commercial considerations can also pose significant challenges to cooperative programs in militarily relevant technologies. For example, the U.S. practice of arms-length relations with private producers, often through the intermediary of a national laboratory, is not mirrored by the practices of some U.S. allies. These governments often work closely with leading national companies, in which the government may also have a direct financial stake, and which have commercial interests in competition with those of U.S. firms.[411] Similarly, U.S.-Japan cooperation in defense and dual-use technology has been asymmetrical. A recent study suggests that, in part, the asymmetry is due to the unwillingness of Japanese industry and government to cooperate technologically on reciprocal terms. The study also suggests that efforts by the U.S. government have lacked consistency and coordination in the pursuit of more balanced technology flows in defense technology

---

[408] Paul Kaminski in *Sources of International Friction and Cooperation in High-Technology Development and Trade.*

[409] Ibid.

[410] Ibid.

[411] Presentation by Rear Admiral Marc Y.E. Pelaez, Office of Naval Research, in *Sources of International Friction and Cooperation in High-Technology Development and Trade.*

collaboration, while U.S. industry has tended to pursue increased sales rather than the acquisition of new technology.[412] This lack of sustained attention to technology may reflect the U.S. preoccupation with the defense of territory during the Cold War, and the tendency—from a position of great economic and technological strength—to subordinate economic interests to diplomatic and military aspects of national security. Japanese policy tends not to make this distinction between national defense objectives and economic interests. Japanese policymakers reject such arguments, seeing technology and production, as well as territory, as national interests that can and must be defended.[413]

These differences in perspective can generate friction with respect both to trade in high-technology products and to cooperative programs to develop new technologies. In some cases, these differences can be outweighed by the substantial benefits offered by equitable international cooperation. However, they underscore the need for care in the establishment and management of cooperative programs. More broadly, as dual-use programs continue or expand in the U.S., Japan, and Europe, increased international friction is likely to occur over program goals and the resources devoted to their achievement.

SUPPLEMENT D.
GLOBAL POSITIONING SYSTEM: GOVERNMENT MISSIONS, COMMERCIAL APPLICATIONS, AND POLICY EVOLUTION

Recent developments in U.S. policy on the Global Positioning System illustrate the powerful synergies made possible by government-funded research and infrastructure investments and by the private sector's development of innovative commercial applications. For national security purposes, the U.S. government invested approximately $10 billion over two decades in a constellation of twenty-four satellites orbiting 11,000 miles above the earth. Each of the satellites has four atomic clocks to provide accurate time. In addition to the satellites, which produce the GPS signals, the "basic" GPS system consists of ground stations, data links, and associated command control facilities. This satellite system is operated and maintained by the U.S. Department of Defense.

A System Developed for Military Advantage—

The initial development of the Global Positioning System was undertaken under the auspices of the U.S. military in the 1970s to provide ex-

---

[412] *Maximizing U.S. Interests in Science and Technology Relations with Japan.* National Academy Press, Washington, D.C., 1995, pp. 4–5.

[413] Richard Samuels, *Rich Nation, Strong Army,* p. 4.

tremely accurate guidance for U.S. aircraft, missiles, and ships anywhere on earth. In combination with other U.S. information-gathering satellites, GPS provides precise targeting data for U.S. weapons systems and accurate positioning information for allied troops. The precision afforded by GPS played a valuable role in U.S. air and ground operations in the course of the Gulf War. Partly on the basis of this experience, the substantial military advantage provided by GPS is now widely recognized, and this innovative system is now being integrated into a wide range of military applications.

—With Growing Civilian Applications

Over the past ten years, however, GPS has evolved far beyond its military origins, becoming "an information resource which supports a wide range of civil, scientific, and commercial functions—from air traffic control to the Internet—through its ability to provide precision location and timing information."[414] Moreover, the rapid advances in microelectronics have contributed to the development of equipment with lower costs, augmented capability, and increased portability.[415] For example, the cost of a receiver has dropped precipitously, from $150,000 per GPS survey receiver in the mid-1980s to approximately $5,000 for an equivalent survey receiver in 1996. GPS handheld receivers for consumers originally sold for approximately $3,000 in 1989, and are now selling at under $200. The GPS engine, or board receiver for the car navigation system, is currently selling, in moderate volume, in the consumer car navigation market at under $100 per unit. By picking up transmissions from the GPS signals and comparing the distances to different GPS satellites, commercial GPS receivers can identify their position on the globe with unprecedented accuracy from 300 meters down to—in some cases—a few centimeters.

While applications are growing rapidly in both the military and commercial sectors, the market for civilian applications now exceeds that for military applications "by roughly three-to-one." This is expected to widen to eight-to-one over the next few years, with projected sales of commercial GPS equipment expected to soar from the current $2 billion to $8.5 billion in the year 2000.[416] The rapid evolution of GPS represents a significant, if

---

[414] Critical Technologies Institute, *A Policy Direction for the Global Positioning System: Balancing National Security and Commercial Interests,* Research Brief, Washington, D.C., December 1995.

[415] A recent Academy report points out that the development of concurrent technologies including satellites and microelectronics was essential to the creation of GPS, as was the assignment of responsibility to the Air Force for the development of a navigation system for all military services. See *Allocating Federal Funds for Science and Technology,* National Research Council, p. 49.

[416] Scott Pace et al., *The Global Positioning System: Assessing National Policies,* Critical Technologies Institute, RAND, Santa Monica, Calif., 1995, p. 201.

specialized, spin-off from national investment in high-technology military systems.[417] It also underscores the unpredictable consequences of technological advance and the powerful synergies which can occur as a result of public investments in new technologies, especially in high-technology infrastructure, and the private sector's development of commercial applications utilizing the innovative capabilities of these systems.[418]

Interestingly, the evolution of the GPS system has benefited from the rapid advance in cost and performance driven by commercial imperatives.[419] While the initial GPS market was derived from government policies designed to achieve defense objectives, the subsequent development of commercial markets has contributed to the ability of the government to accomplish the national security mission of GPS. A recent report notes that "the demand by civilian commercial users of GPS for smaller, better, and cheaper receivers has directly benefited systems designed specifically for military use." The report cites the case of the precision lightweight GPS receivers used by U.S. military forces, which were built at a low cost and delivered on time, in large part because of the technical benefits derived from research and development carried out by private firms for commercial applications.[420]

Issues for U.S. Policy

The rapid growth in commercial applications for the Global Positioning System has required an innovative policy approach to balance U.S. national security objectives with the interests of an increasingly international user community. A recent study noted that the success of the GPS system has raised "complex policy questions for U.S. decision-makers on a variety of issues affecting national defense, commerce, and foreign policy."[421] The report identified four policy areas:

• **Dual-Use issues:** the need to balance the different requirements of national security and commercial interests;

• **Funding regime:** whether the U.S. government should continue to carry the cost of maintaining the Global Positioning System or seek to collect user fees;

• **GPS governance:** how should the Global Positioning System be governed in the future?

---

[417] See Supplement C for a broader discussion of dual-use technologies and commercial spin-offs.

[418] *Allocating Federal Funds for Science and Technology, passim,* especially pp. 31–32.

[419] This process is described in the supplement on dual-use (Supplement C).

[420] Scott Pace et. al., *The Global Positioning System,* p. 251.

[421] Critical Technologies Institute, *A Policy Direction for the Global Positioning System,* p. 1.

- **Concerns of foreign users:** how should the U.S. government guarantee domestic and foreign users—both public and private—that they will have assured, reliable access to GPS signals in a stable policy environment?

These policy areas represent a broad consensus identified in a 1993 report requested by Congress.[422] Asked to assess the future of GPS, the report found that "most aspects of GPS technology, governance and management, and funding are remarkably sound. Among the aspects of the system that are working well and should, therefore, be retained are operational control and funding of the basic GPS satellites by DOD; the aggressive application of GPS technology to public safety and public service needs by civil government agencies; and a dual-use policy that allows room for innovation and entrepreneurship in the GPS industry both at home and abroad." The study emphasized the need for the United States to maintain its leadership position in satellite navigation technology.[423]

Because GPS originated as a military system, national security concerns dominated its initial evolution. As a recent report noted, "the military is necessarily risk-averse in its approach to technology," reflecting its objective of unilateral advantage on the battlefield.[424] These security concerns have, however, been largely overcome by a combination of factors including new technological solutions which should continue to provide a military-competitive edge, the growth in the capabilities of foreign users, and the risks which would be posed by the emergence of competing systems.[425]

As the U.S. security perspectives became less dominant, U.S. commercial interests sought to win acceptance of the view that the successful exploitation of GPS required, above all, a stable policy framework guaranteeing access for all users to the most accurate signals the system could offer. For GPS applications to continue to develop (see below), it is essential the

---

[422] *Legislative Provisions, Global Positioning System.* "In view of the pressures on the defense budget, the necessity for increased civil-military cooperation, the importance of dual-use technology for economic competitiveness and conversion, and the President's interest in effective infrastructure investments, the committee believes that the time has come to examine carefully a number of GPS technical and management issues.

It is clear that GPS offers the potential to revolutionize the movement of people and goods the world over. Civil and commercial exploitation of GPS could soon dwarf that of the Department of Defense and lead to large productivity gains and increased safety in all transportation sectors." *National Defense Authorization Act for Fiscal Year 1994, Conference Report to Accompany H.R. 2401*, Report 103-357, U.S. House of Representatives, Washington, D.C., 10 November 1993, p. 63.

[423] *The Global Positioning System: Charting the Future*, National Academy of Public Administration, Washington, D.C., 1995, p. vii.

[424] Scott Pace et al., *The Global Positioning System.*

[425] Ibid., pp. 186–201. For security-based criticisms of the current U.S. policy, see John Mintz, "U.S. Opens Satellites to Civilians," *The Washington Post*, 30 March 1996.

United States provide a continuous, stable signal, with a widely accepted commitment to avoid the imposition of arbitrary policy changes such as encryption or user fees. A stable policy framework is also seen as the basis for maintaining the current GPS system as a de facto standard. Ultimately, congressional action may be desirable to secure policy stability, an undertaking which should prove possible given the history of broad bipartisan support and common agreement in this area.

The question of standards has enormous stakes, not only in terms of market investments and product development, but also as the system becomes more pervasive, internationally accepted standards become crucial to a wide range of civil, commercial and military applications involving issues such as encryption, safety certification, and international spectrum specifications. A formal, national policy on GPS, providing a predictable environment for both public service and private business decisions, is essential.

A Stable Policy Framework for Investment

From 1983, when President Reagan first offered access to GPS, international reliance on GPS has expanded exponentially. To meet the concerns of increasing reliance on GPS, and the resulting need for a clearly articulated and stable policy framework, in March 1996 the Clinton administration issued a presidential decision directive approving a comprehensive national policy on the future management and use of the U.S. Global Positioning System.[426] The presidential directive was designed to address the broad range of military, civil, commercial, and scientific interests, both national and international, concerned with GPS operations. Its fundamental objective was to establish a clear, high-level commitment to a stable policy environment for the development of international standards facilitating both private and public sector investments. This policy stability is of special interest to public authorities which are increasingly dependent on GPS for the provision of critical public functions.

To this end, the policy guidelines adopted for the operation and management of GPS under this directive included decisions to

• provide the GPS Standard Positioning Service for peaceful, civil, commercial, and scientific use on a continuous, worldwide basis;
• make the service available free of direct user fees;
• discontinue the Selective Availability of GPS signals "within a decade," thereby making available for commercial use the most precise positioning information, formerly reserved only for military applications;

---

[426] See *U.S. Global Positioning System Policy Fact Sheet,* The White House, OSTP/NSC 29 March 1996.

• seek the acceptance of the U.S. GPS as a standard for international use, while ensuring an appropriate balance between international security interests and the requirements of international, civil, commercial, and scientific users;

• establish a permanent GPS Executive Board chaired by the Departments of Defense and Transportation, charged with the management of GPS in consultation with interested parties.

Lessons from GPS

The development of the Global Positioning System captures many of the special features of high-technology development and competition. As noted, the industry grew as a result of U.S. defense expenditure on a military system, although it does not represent the traditional spin-off of defense technologies, insofar as it is the additional capacity of the defense system itself which provides the infrastructure for the industry. A separate GPS satellite signal was made available for commercial use and, as a result, industry investments in R&D created an array of productivity-enhancing information products.

The initial government role was critical to the development of both the technology and the infrastructure. At its origins, the start-up of such a system was most unlikely, not to say impossible, on the part of purely private investors, in terms of both the cost and the untried concept. However, because the government eventually adopted a dual-use approach to the GPS technology, commercial and non-defense public use of the system has expanded far beyond the initial military applications. At the same time, the rapid technical development driven by commercial applications has contributed to increased technical capability and availability for military systems. The military-commercial synergies thus obtained illustrate the power of a dual-use approach to new technologies.

The Growth in Applications

The breadth of benefits made possible by the GPS system were certainly unforeseen when the initial U.S. investment was made. Applications now extend to key public infrastructures such as the Global Information Infrastructure (GII), the Internet, mobile communications and other global wireless applications, and transportation. Public safety applications include civil aviation, disaster management and response, and emergency fire, police, and medical response. The potential applications for private automobiles, and the transport industry generally, are immense. The increasingly widespread use of differential GPS for augmenting basic GPS signals, yields great accuracy, i.e., within centimeters. This capability is translating into "an incredible array of applications," such as demonstrating new systems for landing aircraft in bad weather; robotic plowing, planting, and fertilizing of fields, with tremendous economies and productivity gains; monitoring train and ship locations and even tracking and contributing to the cleanup of oil spills; carrying out twenty-four-hour remote-controlled mining operations; and optimizing the positioning of emergency vehicles to provide the critical response-time margin that saves lives.[427]

The Central Role of International Cooperation

For an information industry to continue to grow globally, however, it requires that national policy evolve into a stable international policy framework. The role of foreign public and private users, and their perceptions and needs, have to be taken into account in U.S. policymaking. As the system evolves toward a de facto global information utility, the need to ensure that the requirements of all stakeholders are accommodated becomes paramount. By seeking regional agreements with Japan and Europe on security and economic issues, the adoption of GPS as a global standard for position location, navigation, and timing can be assured. By pursuing a cooperative approach, the rapid expansion of the industry may continue to be unhindered by potentially debilitating conflict over standards or by the artificial segmentation of global markets.

Finally, notwithstanding these policy challenges, it is important to keep in mind the unprecedented opportunities offered by the Global Positioning System. The growth in GPS applications offers major benefits in terms of consumer welfare, national productivity, and the ability of national, state and local governments to better perform their missions. The evolution of this system highlights the unforeseen benefits to be derived from sustained public investment in new technologies, the benefits of the dual-use approach for military acquisition, and the central role of international cooperation. As information technology becomes increasingly important to international security and commerce, the experience of GPS provides an instructive model for cooperative efforts to promote commercial competition and collective security.

SUPPLEMENT E.
DISCRIMINATORY PUBLIC PROCUREMENT:
PROSPECTS FOR PROGRESS

Of the many tools governments use to pursue their industrial policy goals, one of the oldest and most direct is targeted government procurement. To

---

[427] See *Allocating Federal Funds for Science and Technology,* p. 49.

support noncompetitive domestic companies or to assist in the development of what are expected to be "sunrise" industries, governments often reserve attractive contracts for local suppliers. This can be done through explicit restrictions on who may bid on a contract, or by stated preferences for local suppliers. More often, the government's policy goes unstated, and the mechanism through which it supports local companies is shielded by opaque procurement procedures.

The Government Rationale for Discrimination

In recent years, high-technology industries have been a favorite target of directed government procurement. An experience in Norway in the late 1980s offers a particularly clear example of how some governments have sought to use government procurement to support domestic industry. At that time, the city of Oslo planned to procure an automated toll collection system. After following a competitive bidding procedure, the Oslo authorities decided to sign a contract with a U.S. supplier whose system was cost-competitive and based on proven technology. At that point the Norwegian government— which was providing R&D support to a fledgling Norwegian producer of automated toll collection systems and was also funding the Oslo toll road project— entered the picture and directed Oslo to buy the Norwegian product. Oslo resisted and the matter became the focus of intense media attention. In the process, the Norwegian Minister of Transport stated publicly that if the competition had been between the U.S. firm and another foreign firm, the U.S. firm would have won the contract. However, he continued, in this instance the Norwegian government was obliged to support its domestic supplier because the Oslo contract would give the winner a tremendous boost toward winning similar contracts throughout Europe. It would provide a field test in a European environment, drive down unit costs for the Norwegian producer through economies of scale, create customer acceptance, and influence the development of European standards for toll collection equipment. Ultimately, the Norwegian firm was awarded the Oslo contract and a subsequent contract in the city of Trondheim.[428]

The Power of Government Purchases

The power represented by government purchases is staggering. According to U.S. government estimates, total procurement by central govern-

---

[428] Norway is a member of the Government Procurement Agreement (GPA), and the United States challenged both of these Norwegian procurements in that forum. The U.S. company received compensation for the costs it incurred in bidding on the Oslo project, but was unable to overturn the contract awards.

ments, sub-central governments, and electrical utilities in the European Union is approximately $115 billion per year. Central government procurement in Japan approaches $30 billion. The figures are even more dramatic when the purchasing activities of state-owned enterprises are taken into account. According to a World Bank study, state-owned enterprises accounted for 10.7 percent of economic activity in developing countries and 4.9 percent of economic activity in developed countries from 1978 to 1991.[429] Government procurement policies and decisions are thus one of the most important potential influences on international trade flows.[430]

As a result, the elimination of restrictive procurement practices is a high priority in international trade negotiations. The United States has been especially active. With Japan alone, the United States has negotiated sectoral procurement agreements covering supercomputers, computers, cellular telephones, medical equipment, and telecommunications equipment. As with other trade and investment issues—discussed in the body of this Report—these bilateral initiatives are designed to address discriminatory practices for which effective multilateral arrangements are not in place.

The Exclusion of Government Procurement from Multilateral Disciplines

Although the importance of these purchases was widely recognized, when the international community adopted the General Agreement on Tariffs and Trade in 1948, liberalization of government procurement procedures were not included. In fact, Article III.8 of the GATT specifically excepted public procurement from the principle of national treatment. This exclusion was a

---

[429] Within these statistics, there is considerable variation among countries. During the period 1978 to 1991, state-run enterprises accounted for 10.5 percent of economic activity in France, and 7.1 percent of economic activity in Germany, but only 1.2 percent of economic activity in the United States. *Bureaucrats in Business,* Oxford University Press, New York, 1995, pp. 268–271.

[430] It should be noted, however, that the efficacy of using government procurement to direct business toward domestic high-technology ventures in some sectors is of uncertain value. In the United States, the federal government has succeeded in developing technologies, such as space launch vehicles, where: (1) there was a clear government need and (2) the lack of commercial demand and the high cost of entry made private sector investment uneconomical. The Norwegian toll road case, in which the government sought to provide critical support to launch a new technology, illustrates a third category in which some would argue for the efficacy of government intervention. As a means of supporting new generations of existing commercial technologies, however, the benefits of selling to the government are far less clear. In the semiconductor industry, for instance, government specifications tend to lag behind the state of the art and can impose additional requirements that have no commercial benefit. Companies that become dependent on government contracts may survive, but they are likely to see a steady erosion of their competitive position in the far-larger commercial market.

foregone conclusion from as early as 1946 when the GATT Preparatory Committee recommended in its first session that government procurement *not* be covered by the most-favored-nation (MFN) clause. The participating governments saw their procurement decisions as inherently political and were unwilling to give up the use of one of their most flexible industrial policy tools. This outright exclusion of such a large portion of economic activity was recognized by many as a serious gap in GATT coverage, but for thirty years the gap remained unfilled.

Only in 1979, with the conclusion of the Government Procurement Agreement as one of the Tokyo Round codes, did the GATT contracting parties begin to introduce a measure of discipline and international obligation to government procurement practices. For the purchases it covers, the GPA requires signatories to accord MFN and national treatment to other signatories. It also mandates certain procurement procedures designed to ensure that its objectives are not undercut through the introduction of bid requirements or procedures biased toward a local bidder. The framers of the GPA sought to create an incentive for additional participation by making the benefits of the GPA available only to other signatories.

The GPA, however, also has a number of serious shortcomings:

• First, **participation in the GPA is voluntary**, and the membership from the beginning did not extend essentially beyond OECD countries.[431] As a result, the GPA has done nothing to change procurement practices in the rapidly growing countries which offer great opportunities for increased exports from the industrial countries.[432] The expectation that additional countries would join the GPA in order to obtain the benefits of access to government procurement in the developed world has proven to be unfounded. Most developing countries simply do not expect that their suppliers would be net beneficiaries of the agreement, and under those circumstances they are unwilling to circumscribe their own flexibility in procurement decisions. Limited participation has left procurements in most developing countries open to arbitrary and nontransparent purchasing procedures. Moreover, because transparent procedures offer one of the best defenses against cor-

---

[431] The following countries currently apply the GPA: Canada; the European Union (including Austria, Belgium, Denmark, Finland, France, Germany, Greece, Ireland, Italy, Luxembourg, the Netherlands, Portugal, Spain, Sweden, and the United Kingdom); Israel; Japan; Norway; Switzerland; and the United States. Korea will begin to apply the GPA on January 1, 1997. Singapore is in the process of acceding to the agreement.

[432] U.S. exports to the ten "Big Emerging Markets" (BEMs) identified by the U.S. Commerce Department already exceed U.S. exports to either Japan or the European Union, and by the year 2000 they are expected to exceed U.S. exports to Japan and the EU combined. None of these eighteen countries now apply the GPA, and only Korea and Singapore are likely to begin applying the agreement in the near future.

rupt behavior in government procurement, failure to sign up more members to the GPA has meant that one of the great potential values of the GPA has gone unrealized.

• Second, **the GPA disciplines apply only to procuring entities "scheduled" by the signatories.** Although the United States scheduled most of its central government procurement in 1979, "entity coverage" by other countries was more limited, resulting in significant imbalances. For instance, prior to the Uruguay Round agreement, the European Union and most other signatories refused to schedule their telecommunications, electrical, transportation, and water utilities. In fact, very few countries scheduled their state-owned enterprises, whatever their field of business.[433] U.S. companies operating in these sectors thus obtained no new market access under the GPA. The imbalance was compounded by the fact that shortly after the GPA went into effect the United States introduced competition—including competition from international suppliers—into its own telecommunications and electrical utility markets. U.S. dissatisfaction with this imbalance, combined with European desires to obtain more U.S. coverage of state and local entities, resulted in a serious U.S.-EU trade dispute. This dispute led to agreements in 1993 and 1994 to undertake coverage of electrical utilities and subcentral entities,[434] but telecommunications procurement issues were not resolved, resulting in retaliation by both sides.

• Third, **the GPA procedures are difficult to understand.** While they are laudable in their intentions, the excessive detail of the GPA procedures, particularly by comparison with other elements of the GATT, tends to wrap the GPA in a cloak of mystery. It is only a select few, the high priests of the agreement as it were, who understand the commitments and undertake to interpret them for the rest of the world. Even if they were to feel that GPA participation were otherwise warranted, this factor deters nonsignatories

---

[433] State trading companies are covered by Article XVII of the GATT, which requires non-discrimination in their purchasing practices, but exempts purchases that are ultimately intended for governmental purposes.

[434] Coverage of electrical utilities and U.S. subcentral entities was achieved first through a bilateral agreement, and, after 1 January 1995, through coverage of most of these entities in the GPA. The bilateral agreement did not, however, prove satisfactory. For example, U.S. suppliers of heavy electrical equipment claim that they continued to encounter discriminatory procurement procedures and inadequate remedies in the German market. As a result, on 30 April 1996 the U.S. Trade Representative identified Germany (under Title VII of the 1988 Omnibus Trade and Competitiveness Act) for "a significant and persistent pattern" of discrimination and failure to adequately implement (its government procurement) obligations." This identification of Germany triggered a sixty-day period of consultations including, at this writing, the possibility of retaliatory action by the U.S. government if the consultations do not prove successful. Whatever the eventual resolution of this dispute, it highlights both the importance of high-technology markets to even the most industrialized countries and the persistence of government procurement of high-technology products as a source of friction within the international system.

from participation, to the extent their authorities believe they are dependent on foreign expert interpretations of the agreement and do not wish to make themselves vulnerable to arcane dispute settlement procedures.

• Fourth, **the GPA is based on an all or nothing principle that differs from the rest of the GATT.** If an entity is scheduled and the GPA is applied for that entity to another GPA signatory, the entity is subject to all GPA disciplines. If not, it is subject to no disciplines. This all-or-nothing approach leaves no room for the type of steady improvement in market access that has been so successful with tariff reductions in the GATT.

Since its adoption, the Government Procurement Agreement has been modified through almost continuous negotiations, but GPA membership has expanded only marginally.[435] In an attempt to plug procedural holes that permitted signatories to follow the GPA rules and still steer contracts to local suppliers, most modifications to the GPA during the 1980s added new procedural requirements, further complicating the agreement.

Progress in the Uruguay Round

During the Uruguay Round, substantial progress was made on several government procurement fronts.[436] For the first time, countries were permitted to take derogations from their national treatment obligations, thus moving a step away from the all-or-nothing principle. The GPA was also expanded to cover the procurement of services (including construction services) and subcentral entities and public utilities. When combined with additional entity coverage at the central government level, these changes resulted in a tenfold increase in the value of procurements covered by GPA procedures.[437] Another major addition to the GPA was a requirement that signatories provide procedures for bid challenges as well as judicial remedies in disputes over the award of a contract. The GPA thus serves a valuable purpose. It has shown that governments can subject themselves to international obligations in the field of government procurement, and it has opened hundreds of billions of dollars of contracts to international competition.

None of these changes, however, adequately addresses the fundamental flaws with the GPA as identified above. The prospects for any significant expansion in membership remain limited;[438] entity coverage among existing

---

[435] Most of the growth in GPA membership has occurred through enlargement of the European Union. Greece, Spain, and Portugal all joined the GPA after joining the EU.

[436] Technically, the government procurement negotiations, which were ongoing at the time the Uruguay Round began, were not part of the Round. In practice, however, revisions to the GPA made up an important part of the final Uruguay Round package.

[437] "Creating open competition in government procurement," WTO Internet Home Page, 1996.

[438] Unlike most of the other Tokyo Round codes, the GPA was not incorporated as an integral element of the Uruguay Round single act, and membership remains voluntary.

signatories is still incomplete; procurement procedures remain unnecessarily intimidating, and the structure of the agreement provides slim hopes for achieving improvements outside of the existing membership. As a result, some analysts believe the time has come for a new approach; an approach that will recognize and lock in the progress made to date by existing GPA members, demystify the agreement, and encourage a steady process of liberalization rather than expecting new GPA members to leap directly to full liberalization.[439] In doing so, an expanded GPA would expose procurement practices to public scrutiny, thereby contributing to the worldwide effort to reduce corruption in government. A broader, more effective agreement would also offer a means to reduce a major source of friction in high-technology trade while providing the benefits of transparent competition to government acquisitions of high-technology products.

A New GPA

A new Government Procurement Agreement would look to the GATT itself, and in particular the success of the GATT in eliminating quotas and reducing tariffs, as a model for steady improvements in trade liberalization.[440] With this model in mind, a reconstituted agreement might embody the following six elements:

1. **Membership:** all WTO members would be required to join the GPA.
2. **Entity coverage:** GPA members would be required to make binding commitments covering purchases by all of their government entities, including all state-owned enterprises. Under a reconstituted GPA, unlike the existing one, however, these commitments could involve exceptions from national treatment, such as a specified preference level for domestic suppliers. Existing GPA members should, at a minimum, bind their existing commitments with respect to other current GPA signatories (including the

---

[439] See the statement by R. Michael Gadbaw in *Sources of International Friction and Cooperation in High-Technology Development and Trade.*

[440] At the first GATT/ITO Preparatory Committee meeting in London in 1946, the United States set out five objectives for the new organization, many of which could be applied today to the GPA. Professor John Jackson describes these objectives as follows: "(1) Existing barriers to international trade should be substantially reduced; (2) International trade should be multilateral rather than bilateral; (3) International trade should be non-discriminatory; (4) Stabilization policies for industry and agriculture...[are] so intimately related to international trade policies that the two must be consistent and coordinated; and (5) The rules for international commerce should be drafted so that they would apply with equal fairness and equal force to the external trade of all nations regardless of whether their internal economies were organized upon the basis of individualism, collectivism or some combination of the two." John H. Jackson, *World Trade and the Law of the GATT,* Bobbs-Merrill Company, New York, p. 54.

national treatment element inherent in those commitments). Transparency (item 4 below) and remedy (item 5 below) procedures should apply to all entities in all WTO members.

3. **Liberalization method:** GPA members would engage in regular negotiations to improve their degree of procurement liberalization. In these negotiations, as in tariff negotiations, countries could trade off reductions in their discriminatory practices in exchange for similar reductions by others.

4. **Procedures:** The GPA would be amended to focus on the key commitments necessary to achieve improved procurement opportunities. These include requirements for: (a) adequate notice; (b) neutral standards (i.e., standards should not be set in a way that gives an advantage to any one bidder); (c) objective and pre-specified bid criteria; (d) public bid opening; and (e) award of contracts to the lowest compliant bidder or demonstrably best overall value on the basis of the objective criteria. The essential elements of these commitments should be distilled and inserted into the GPA so that the principles are readily apparent and readily applied.

5. **Remedies:** The current procedures on bid challenge, access to an independent appeal body, and dispute settlement should be retained. In addition, once the GPA becomes a requirement for all WTO signatories, cross-retaliation[441] should be permitted. In the event that a WTO member were found by a panel to have violated the GPA but did not implement the panel ruling, cross-retaliation would increase the options for retaliation available to an aggrieved party. This should provide an improved deterrent to violations of the agreement.

6. **National Treatment/MFN:** To accord with the shift from an all-or-nothing approach to an approach that seeks broad and steady liberalization, the national treatment requirement of the GPA would have to be eliminated, although the MFN requirement should be retained as a core principle.

Current Prospects

Leading WTO members have begun to consider the need for changes to the GPA. At the meeting in Kobe, Japan, on 1921 April 1996, the trade ministers of Japan, Canada, the European Union, and the United States (the Quad ministers) addressed the need to bring additional members into the GPA and endorsed an approach other than the all-or-nothing approach to procurement liberalization. The chairman's statement from that meeting

---

[441] Cross-retaliation would permit retaliation outside of the government procurement area if a GPA member were to violate its GPA commitments and then fail to implement a panel ruling against it. Full cross-retaliation would also permit retaliation in the government procurement area for violations related to other WTO agreements. Currently, the GPA does not permit cross-retaliation.

reports, "We agreed to renew our efforts to expand membership in the WTO Agreement on Government Procurement (GPA) and to improve its disciplines through reducing barriers to government procurement. As a first step, we agreed to initiate work on an interim arrangement on transparency, openness and due process in government procurement, which would help to reduce corruption as an impediment to trade."[442]

While the Quad ministers did not provide any further detail on what an interim agreement on transparency, openness, and due process might be, the market-opening potential of such an agreement is enormous. If an interim agreement included the procedures described in points (4) and (5) above, it would offer three principal benefits. First, such an agreement would serve as a means by which significant numbers of additional countries could sign on to government procurement disciplines. Second, even without a national treatment commitment, transparency provisions alone would greatly improve the environment for foreign companies seeking to do business in signatories to the interim agreement. Third, as the Quad ministers' statement suggests, a truly transparent procurement system is incompatible with illicit payments. For all of these reasons, the transparency initiative represents a welcome step.

In the short term, however, the road to adoption of an interim agreement with provisions on transparency, openness, and due process is likely to prove difficult. A recent statement by the ASEAN economic and trade ministers opposed discussion of corruption issues (including the issue of transparency in procurement) at the Singapore Ministerial meeting. This is of particular concern because Singapore, the host of the Ministerial meeting, is an ASEAN member.

The position taken at the meeting of the ASEAN economic and trade ministers highlights the different perspectives of countries at different levels of development. As described in the first section of the report, it also illustrates the different policy objectives of competitors for high-technology industry in the global economy. And at a more fundamental level, it may be the ASEAN position reflects differing values and a determination on the part of newly industrialized countries to assert their independence from Western precepts.

Responding to this ASEAN challenge will require a clear presentation of the benefits that can accrue to those countries participating in efforts to increase transparency. In this regard, the Quad ministers' conclusion that transparency, broader participation in the Government Procurement Agreement, and improvement of its disciplines is a positive step. For further progress, it is likely to be necessary to undertake the fundamental reform of the Government Procurement Agreement outlined above.

---

[442] Chairman's Statement, Kobe Quadrilateral Trade Ministers' Meeting 1921 April 1996, p. 2.

Bibliography

Economic Report of the President, U.S. Government Printing Office, Washington, D.C., February 1991.

Trans-Atlantic Business Dialogue: Overall Conclusions, 11 November 1995. Seville, Spain.

U.S. Business Access to Certain Foreign State of the Art Technology, GAS/NSIAD-91-278. U.S. Government Printing Office, Washington, D.C., September 1991.

"U.S./China Make Progress in Talks; Many Industries Urge U.S. Action." *International Trade Reporter,* vol. 12, no. 4, 25 January 1995.

Alic, John, et al. (eds.), *Beyond Spinoff: Military and Commercial Technologies in a Changing World.* Harvard Business School Press, Boston, Mass., 1992.

Amsden, Alice H., *Asia's Next Giant: South Korea and Late Industrialization.* Oxford University Press, New York, 1989.

Amsden, Alice H., *Diffusion of Development: The Late Industrializing Model and Greater East Asia.* Papers and Proceedings of the American Economic Association, May 1991.

Bailey, David, George Harte, and Roger Sugden, "U.S. Policy Debate towards Inward Investment," *Journal of World Trade,* vol. 26, August 1992.

Baily, Martin N., and A. Chakrabarti, *Innovation and the Productivity Crisis.* The Brookings Institution, Washington, D.C., 1988.

Baldwin, Richard E. Jr., *The Role of Government Procurement.* 21 November 1995 contribution to Steering Committee deliberations

Baldwin, Richard E. (ed.), *Trade Policy Issues and Empirical Analysis.* University of Chicago Press, Chicago, 1988.

Barber, Randy and Robert E. Scott, *Jobs on the Wing: Trading Away the Future of the U.S. Aerospace Industry.* Economic Policy Institute, Washington, D.C., 1995.

Blecker, Robert A., *Beyond the Twin Deficits.* M.E. Sharpe, New York, 1992.

Blustein, Paul, and Steve Mufson, "A China Trade Question: Is it Ready for Rules?" *The Washington Post,* 19 May 1996.

Boltuck, Richard, and Robert Litan (eds.), *Down in the Dumps: Administration of the Unfair Trade Laws.* The Brookings Institution, Washington, D.C., 1991.

Borrus, Michael, and Jeffrey A. Hart, "Display's the Thing: The Real Stakes in the Conflict over High-Resolution Displays," *Journal of Policy Analysis and Management,* vol. 13, no. 1, 1994.

Borrus, Michael, Wayne Sandholtz, John Zysman, Ken Conca, Jay Stowsky, Steven Vogel, and Steve Weber, *The Highest Stakes: The Economic*

Foundations of the Next Security System. Oxford University Press, New York, 1992.

Bozeman, B., A. Link, and A. Zardhoohi, "An Economic Analysis of R&D Joint Ventures," *Management and Decision Economics,* 1986.

Brander, James, and Barbara Spencer, "Export Subsidies and International Market Share Rivalry," *Journal of International Economics,* February 1985

Brody, Richard J., *Effective Partnering: A Report to Congress on Federal Technology Programs.* Office of Technology Policy, U.S. Department of Commerce, Washington, D.C., April 1996.

Brown, Martin, *Impacts of National Technology Programs,* OECD, Paris, 1995.

Bulletin of the European Union, *Green Paper on innovation,* Supplement 5/95, Office for Official Publications of the European Communities, Luxembourg, 1996.

Bureau of Export Administration, *Offsets in Defense Trade.* U.S. Department of Commerce, Washington, D.C., May 1996.

Bureau of Export Aministration Public Affairs Office, BXA-96-8, Washington, D.C., 17 May 1996.

Cadot, Olivier, H. Landis Gabel, Jonathan Story and Douglas Webber, *European Casebook on Industrial and Trade Policy.* Prentice Hall, New York, 1996.

Carnoy, Martin, Manuel Castells, Stephen S. Cohen and Fernando Henrique Cardoso, *The New Global Economy in the Information Age: Reflections on Our Changing World.* The Pennsylvania State University, University Park, Pa., 1993.

Chang, Ha-Joon, *The Political Economy of Industrial Policy.* St. Martin's Press, New York, 1994.

Chang, Ha-Joon, and Robert Rowthorn, (eds.), *The Role of the State in Economic Change.* Clarendon Press, Oxford, U.K., 1995.

Chimerine, Lawrence, "U.S. Trade Deficits: Causes and Policy Implications." Economic Strategy Institute, Washington, D.C., 1995.

Coalition for Intelligent Manufacturing Systems and the Office of Technology Policy, Technology Administration, U.S. Department of Commerce *Intelligent Manufacturing Systems.* Washington, D.C., 1995.

Cohen, Linda, and Roger G. Noll, *The Technology Pork Barrel.* The Brookings Institution, Washington, D.C., 1991.

Cohen, Stephen, and Pei-Hsiung Chin, *Tipping the Balance: Trade Conflicts and the Necessity of Managed Competition in Strategic Industries.* Kiel Institute of World Economics, *Towards a New Global Framework for High-Technology Competition, 30–31 August 1995 Conference Proceedings.* Kiel, Germany, forthcoming.

Council on Competitiveness, *Roadmap for Results: Trade Policy, Technology and American Competitiveness.* Washington, D.C., 1993.

Council of Economic Advisers, *Supporting Research and Development to Promote Economic Growth: The Federal Government's Role.* Washington, D.C., October 1995.

Critical Technologies Institute, *A Policy Direction for the Global Positioning System: Balancing National Security and Commercial Interests.* RAND Research Brief. Washington, D.C., December 1995.

Dasgupta, Partha, and Paul A. David, "Toward a New Economics of Science," *Research Policy,* vol. 23, 1994.

David, Paul A., and Dominique Foray, *STI Review.* OECD, Paris, 1995.

David, Paul A., David C. Mowery, and W. Edward Steinmueller, "Government-Industry Research Collaborations: Managing Missions in Conflict," paper presented at CEPR/AAAS conference *University Goals, Institutional Mechanisms, and the 'Industrial Transferability' of Research,* Stanford, Calif., 18–20 March 1994.

Defense Science Board Industrial Base Committee, *Report of the Defense Science Board Task Force: Foreign Ownership and Control of U.S. Industry.* Washington, D.C., June 1990.

Denham, Robert, and Michael Porter, *Lifting All Boats: Increasing the Payoff in the Private Investment of the U.S. Economy.* Report of the Capital Allocation Sub-Council to the Competitiveness Policy Council, Washington, D.C., September 1995.

Department of Commerce, *Foreign Direct Investment in the United States: An Update: Review and Analysis of Current Developments,* Washington, D.C., 1993.

Dumas, Lloyd J., *The Overburdened Economy.* University of California Press, Berkeley, Calif., 1986.

Eaton, Jonathan and Samuel Kortum, "Trade in Ideas: Patenting and Productivity in the OECD," *Journal of International Economics,* 1996 forthcoming.

Economic Strategy Institute, *China and the WTO: Economy at the Crossroads,* Washington, D.C., November 1994.

Electronic Industry Association of Japan, *Mission Accomplished: Why There Is No Need for a Semiconductor Arrangement with Japan.* Tokyo, January 1996.

Encarnation, Dennis J., *Rivals Beyond Trade: America versus Japan in Global Trade.* Cornell University Press, Ithaca, 1992.

Etzoini, Amitai, *The Moon-Doggle.* Doubleday, New York, 1964.

European Commission, Directorate-General XIII, Telecommunications, Information Market and Exploitation of Research, *The European Report for Science and Technology Indicators, 1994.* Luxembourg, October 1994.

European Commission, *Research and Technology: the Fourth Framework Programme (1994-1998).* Brussels, Belgium, 1995.

European Commission for Science, Research, and Development, *RTD Info.* Brussels, February 1996.

Fallows, James, *Looking at the Sun: The Rise of the New East Asian Economic and Political System.* Pantheon Books, New York, 1994.

Fallows, James, *National Defense.* Random House, New York, 1981.

Fettig, David (ed.), *The Economic War among the States.* Federal Reserve Bank of Minneapolis, Minneapolis, Minn., May 1996.

Fields, Karl J., *Enterprise and the State in Korea and Taiwan.* Cornell University Press, Ithaca, N.Y., 1995.

Fine, Charles H., and Daniel E. Whitney, "Is the Make-Buy Decision a Core Competence?" MIT Center for Technology, Policy, and Industrial Development, Cambridge, Mass., January 1996.

Flamm, Kenneth, *Mismanaged Trade? Strategic Policy and the Semiconductor Industry.* The Brookings Institution, Washington, D.C., 1996.

Flat Panel Display Task Force, *Building U.S. Capabilities in Flat Panel Displays, Final Report,* U.S. Department of Defense, Washington, D.C., October 1994.

Florida, Richard, *International Investment: Neglected Engine of the Global Economy.* American Enterprise Institute, Washington, D.C., 1994.

Fong, Glenn R., "State Strength, Industry Structure, and Industrial Policy: American and Japanese Experiences in Microelectronics," *Comparative Politics,* vol. 22, no. 3, April 1990.

Friedman, David and Richard Samuels, *How to Succeed Without Really Flying: The Japanese Aircraft Industry and Japan's Technology Ideology.* MIT Japan Program Working Paper 92-01, presented at NBER Conference on Japan and the U.S. in Pacific Asia, San Diego, Calif., April 1992.

Gansler, Jacques, *Defense Conversion, Transforming the Arsenal of Democracy,* MIT Press, Cambridge, Mass., 1995.

Gansler, Jacques, "Integrating Civilian and Military Industry," *Issues in Science and Technology.* Fall, 1988.

Garten, Jeffrey A., *A Cold Peace: America, Japan, Germany, and the Struggle for Supremacy.* The Twentieth Century Fund, New York, 1992.

Gaster, Robin, and Clyde V. Prestowitz Jr., *Shrinking the Atlantic: Europe and the American Economy.* North Atlantic Research Inc. and Economic Strategy Institute, Washington, D.C., June 1994.

Gaster, Robin, "Guiding Foreign Investment," *Foreign Policy,* Fall 1992.

General Accounting Office, *Analyzing National Security Concerns,* GAO/NSIAD-90-94, Washington, D.C., 1990.

General Accounting Office, *SEMATECH's Technological Progress and Proposed R&D Program,* GAO/RCEED-92-223 BR, Washington, D.C., July 1992.

Gerschenkron, A., *Economic Backwardness in Historical Perspective.* Harvard University Press, Cambridge, Mass., 1962.

Glain, Steve, "Concern Over 'Menace' Dissipates as Japan, U.S. Unveil Fighter Jet," *Wall Street Journal*, 22 March 1996.

Graham, Edward, and Paul Krugman, *Foreign Direct Investment in the United States*. Institute for International Economics, Washington, D.C., 1991 and 1995.

Griliches, Zvi, *The Search for R&D Spillovers*. Harvard University, Cambridge, Mass., 1990.

Grindley, Peter, David C. Mowery, and Brian Silverman, "SEMATECH and Collaborative Research: Lessons in the Design of High-Technology Consortia," *Journal of Policy Analysis and Management*, vol. 13, no. 4, 1996.

Ham, Rose Marie, and David Mowery, "Improving Industry-Government Cooperative R&D, *Issues in Science and Technology*, Summer 1995.

Hart, Jeffrey A., "Anti-Dumping Petition of the Advanced Display Manufacturers of America: Origins and Consequences." Paper delivered at Annual Meeting of the International Studies Association, Atlanta, Ga., 1–4 April 1992.

Hart, Jeffrey A., and Aseem Prakash, "Implications of Strategic Trade for the World Economic Order," Paper prepared for the Annual Meeting of the International Studies Association, San Diego, Calif. 16–20 April 1996.

Hart, Jeffrey A., *Rival Capitalists: International Competitiveness in the United States, Japan, and Western Europe*. Cornell University Press, Ithaca, N.Y., 1992.

Hindley, B. (ed.), *State Investment Companies in Western Europe*. Macmillan Press, London, 1983.

Hitgartner, S., and S.I. Brandt-Rauf, "Controlling Data and Resources: Access Strategies in Molecular Genetics." Paper presented at the CEPR/ AAAS Conference *University Goals, Institutional Mechanisms and the 'Industrial Transferability' of Research*, Stanford, Calif., 18–20 March 1994.

Hoekman, Bernard M., and Petros C. Mavroidis, *Policy Externalities and High-Tech Rivalry: Competition and Multilateral Cooperation Beyond the WTO*. OECD, Paris, October 1995.

Howell, Thomas, "Dumping: Still a Problem in International Trade," *Trade and Competition Policies*. Westview Press, Boulder, Colo., forthcoming.

Howell, Thomas, Brent Bartlett, and Warren Davis, *Creating Advantage: Semiconductors and Government Industrial Policy in the 1990s*. SIA, Santa Clara, Calif., 1992.

Howell, Thomas R., Jeffrey D. Nuechterlein, and Susan B. Hester, *Semiconductors in China: Defining American Interests*. Semiconductor Industries Association, Dewey Ballantine, Washington, D.C., 1995.

Huber, Thomas M., *Strategic Economy in Japan.* Westview Press, San Francisco, Calif., 1994.

International Trade Administration, *1993 U.S. Industrial Outlook,* U.S. Department of Commerce, Washington, D.C.

Irwin, Douglas A. and Peter J. Klenow, *SEMATECH: Purpose and Performance.* Paper Prepared for the NAS Colloquium on Science, Technology and the Economy, Irvine, Calif., 20–22 October 1995.

Jackson, John H., *The World Trading System.* MIT Press, Cambridge, Mass., 1989.

Johnson, Chalmers, *MITI and the Japanese Economic Miracle: The Growth of Industrial Policy, 1925–1975.* Stanford University Press, Stanford, Calif., 1982.

Kaldor, Mary, *The Baroque Arsenal.* Hill and Wang, New York, 1981.

Kaltenheuser, Skip, Commentary, *New Technology Week,* 5 February 1996.

Katz, M.L., and J.A. Ordover, "R&D Cooperation and Competition," *Brookings Papers on Economic Activity,* The Brookings Institution, Washington, D.C., 1990.

Kende, Michael, "Government Support of the European Information Technology Industry." Prepared for the CEPR/WZB Conference *Does Europe Have an Industrial Policy,* Berlin, 19–20 April 1996.

Koopman, Georg, and Hans-Eckart Scharrer (eds.), *The Economics of High-Technology Competition and Cooperation in Global Markets.* HWWA Institute for Economic Research, Hamburg, Germany, 1996.

Korb, Lawrence J., "Military Metamorphosis," *Issues in Science and Technology.* Winter 1995–1996.

Krugman, Paul, *Peddling Prosperity.* W.W. Norton Press, New York, 1994.

Landau, Ralph and Nathan Rosenberg (eds.), *The Positive Sum Strategy: Harnessing Technology for Economic Growth.* National Academy Press, Washington, D.C., 1986.

Landau, Ralph, Timothy Taylor, and Gavin Wright, *The Mosaic of Economic Growth.* Stanford University Press, Stanford, Calif., 1996.

Lawrence, Robert Z., "Japan's Low Levels of Inward Investment: The Role of Inhibitions on Acquisitions," *Transnational Corporations,* vol. 1, no. 3, December 1992.

List, Friedrich, *The National System of Political Economy,* translation by Sampson S. Lloyd, Augustus M. Kelly Publisher, New York, 1966.

Maejima, Susumu, "Ending Chip Pact Is Shortsighted..." *Asahi Evening News,* 5 February 1996.

Mansfield, Edwin, "Academic Research and Industrial Innovation," *Research Policy,* February 1991.

Mason, Mark, *American Multinationals and Japan: The Political Economy of Japanese Capital Controls.* Harvard University Press, Cambridge, Mass., 1992.

Melman, Seymour (ed.), *The War Economy of the United States: Readings on Military Industry and Economy.* St. Martin's Press, New York, 1971.

Milgrom, Paul, and J. Roberts, "Limit Pricing under Incomplete Information," *Econometrica,* vol. 50, no. 2, March 1982.

Ministry of Foreign Affairs, Japan, *Salient Points and Data Related to the Japan-U.S. Semiconductor Arrangement: Demonstrated Success in Achieving Results Through Free Trade.* Tokyo, February 1996.

Mintz, John, "U.S. Opens Satellites to Civilians," *The Washington Post,* 30 March 1996.

Mowery, David C., and Nathan Rosenberg, *The Japanese Commercial Aircraft Industry since 1945: Government Policy, Technical Development, and Industrial Structure.* The International Strategic Institute at Stanford, Stanford, Calif., 1985.

Mowery, David C., and Nathan Rosenberg, *Technology and the Pursuit of Economic Growth.* Cambridge University Press, New York, 1989.

Mowery, David C., and Nathan Rosenberg, "New Developments in U.S. Technology Policy: Implications of Competitiveness and International Trade Policy," *California Management Review,* vol. 32, no. 1, Fall 1989.

Murphy, John J., and Paula Stern, *A Trade Policy for a More Competitive America.* Report of the Trade Policy Subcouncil to the Competitiveness Policy Council, Washington, D.C., March 1993.

Nadiri, Ishaq, *Innovations and Technological Spillovers.* NBER Working Paper No. 4423, Cambridge, Mass., 1993.

National Research Council, *Allocating Federal Funds for Science and Technology.* National Academy Press, Washington, D.C., 1995.

National Research Council, *High-Stakes Aviation: U.S.-Japan Technology Linkages in Transport Aircraft.* National Academy Press, Washington, D.C., 1994.

National Research Council, *Maximizing U.S. Interests in Science and Technology Relations with Japan: Report of the Defense Task Force.* National Academy Press, Washington, D.C., November 1995.

National Research Council, *Review of the Research Program of the Partnership for a New Generation of Vehicles (PNGV).* National Academy Press, Washington, D.C., 1994.

National Research Council, *Standards, Conformity Assessment, and Trade into the 21st Century.* National Academy Press, Washington, D.C., 1995.

National Research Council, *Technology, Wages, Productivity, and Employment, 1-2 May 1995, Conference Proceedings.* National Academy Press, Washington, D.C., forthcoming.

National Research Council, *The Unpredictable Certainty: Information Infrastructure Through 2000.* National Academy Press, Washington, D.C., 1996.

National Research Council, *U.S.-Japan Strategic Alliances in the Semiconductor Industry: Technology Transfer, Competition, and Public Policy.* National Academy Press, Washington, D.C., 1995.

National Science Board, *Science and Engineering Indicators—1993.* U.S. Government Printing Office, Washington, D.C., 1993.

National Science Board, *Science and Engineering Indicators—1996.* U.S. Government Printing Office, Washington, D.C., 1996.

Nelson, Richard (ed.), *Government and Technical Progress: A Cross Industry Analysis.* Pergamon, New York, 1982.

Nelson, Richard R. (ed.), *National Innovation Systems: A Comparative Study.* Oxford University Press, New York, 1993.

Neven, Damien, and Paul Seabright, "European Industrial Policy: The Airbus Case," *Economic Policy,* no. 21, October 1995.

Nohria, Nitin, and Robert G. Eccles (eds.), *Networks and Organizations: Structure, Form, and Action.* Harvard Business School Press, Boston, Mass., 1992.

North, Douglass, "Economic Performance through Time," Nobel Prize acceptance speech, December 1993, as reprinted in *The American Economic Review,* June 1994.

OECD, *Industrial Subsidies: A Reporting Manual.* Paris, 1995.

OECD, *Market Access and Competition in Technology-Intensive Industries: Issues Paper.* Paris, 26-27 October 1995.

OECD, *Meeting of the Committee for Scientific and Technological Policy at the Ministerial Level: International Technology Cooperation, Discussion Paper, Communique, and Recommendation,* Directorate for Science, Technology and Industry, Paris, September 1995.

OECD, *Ministerial Communiqué,* SG/PRESS (95) 41, Paris, 24 May 1995.

OECD, *Recommendations of the Council Concerning Principles for Facilitating International Cooperation Involving Enterprises.* Paris, 27 September 1995.

Office of Technology Assessment, *Arming Our Allies: Cooperation and Competition in Defense Technology.* Congress of the United States, Washington, D.C., May 1990.

Office of Technology Assessment, *Competing Economies: America, Europe, and the Pacific Rim.* Congress of the United States, Washington, D.C., October 1991.

Office of Technology Assessment, *International Partnerships in Large Science Projects.* Congress of the United States, July 1995.

Office of Technology Assessment, *Multinationals and the National Interest: Playing by Different Rules.* Congress of the United States, 1993.

Office of Technology Assessment, *Multinationals and the U.S. Technology Base.* Congress of the United States, Washington, D.C., September 1994.

Office of Science and Technology Policy, *U.S. Global Positioning System Policy Fact Sheet.* The White House, Washington, D.C., 29 March 1996.

Office of Technology Policy, *Globalizing Industrial Research and Development.* U.S. Department of Commerce, Washington, D.C., 1995.

Okimoto, Daniel I., *Between MITI and the Market: Japanese Industrial Policy for High-Technology.* Stanford University Press, Stanford, Calif., 1989.

Ostry, Sylvia, and Richard R. Nelson, *Techno-Nationalism and Techno-Globalism: Conflict and Cooperation.* The Brookings Institution, Washington, D.C., 1994.

Ostry, Sylvia, "The Post Uruguay Trading System: The Major Challenges," Industry Canada Distinguished Speakers Series, Ottawa. 12 May 1995

Ostry, Sylvia, "Technology Issues in the International Trading System." OECD, Paris, 26–27 October 1995 and in Charles W. Wessner (ed.), *Sources of International Friction and Cooperation in High-Technology Development and Trade, 30–31 May 1995, Papers and Proceedings.* National Academy Press, Washington, D.C., forthcoming.

Pace, Scott et al., *The Global Positioning System: Assessing National Policies.* Critical Technologies Institute, RAND, Santa Monica, Calif., 1995.

Patel, Pari, and Keith Pavitt, "Large Firms in the Production of the World's Technology: An Important Case of Non-Globalization," *Journal of International Business Studies,* First Quarter 1991.

Pavitt, Keith, "National Policies for Technical Change: Where Are There Increasing Returns to Economic Research?" Paper presented at the Colloquium on *Science, Technology and the Economy,* 20–22 August 1995, Irvine, Calif.

Perret, Geoffrey, *A Country Made by War: From the Revolution to Vietnam— The Story of America's Rise to Power.* Random House, New York, 1989.

Porter, Michael, *The Competitive Advantage of Nations.* Free Press, New York, 1990.

Prestowitz, Clyde, *Trading Places.* Basic Books, New York, 1988.

Price, Daniel M., *Investment Rules and High-Technology: Toward a Multilateral Agreement on Investment.* OECD, Paris, France, 26-27 October 1995.

Procassini, Andrew A., *Competitors in Alliance: Industry Associations, Global Rivalries, and Business-Government Relations.* Quorum Books, Westport, Conn., 1995.

Rashish, Peter S. (ed.), *Building Blocks for Transatlantic Economic Area.* European Institute, Washington, D.C., 1996.

Rausch, Lawrence M., *Asia's New High-Tech Competitors,* NSF 95-309. National Science Foundation, Arlington, Va., 1995.

Reich, Simon, "Asymmetries in National Patterns of Foreign Direct Investment: Consequences for Trade and Technology Development," in Charles

W. Wessner (ed.), *Sources of International Friction and Cooperation in High-Technology Development and Trade, 30–31 May 1995, Papers and Proceedings.* National Academy Press, Washington, D.C., forthcoming.

Reid, Proctor P., and Alan Schriesheim (eds.) *Foreign Participation in U.S. Research and Development: Asset or Liability?* National Academy of Engineering, National Academy Press, Washington, D.C., 1996.

Rembser, Josef, *Intergovernmental and International Consultations/Agreements and Legal Co-operation Mechanisms in Megascience: Experiences, Aspects, and Ideas.* OECD, Paris, 1995.

Rosen, Stephen, *Winning the Next War: Innovation and the Modern Military.* Cornell University Press, Ithaca, N.Y., 1991.

Rosenberg, Nathan, "Civilian 'Spillovers' from Military R&D Spending: The U.S. Experience Since World War II," in S. Lakoff and R. Willoughby (eds.), *Strategic Defense and the Western Alliance.* Lexington Books, Lexington, Mass., 1987.

Rosenbloom, Richard S., and William J. Spencer, *Engines of Innovation: U.S. Industrial Research at the End of an Era.* Harvard Business School Press, Boston, Mass., 1996.

Rosenbloom, Richard S., and William J. Spencer, "The Transformation of Industrial Research." *Issues in Science and Technology,* Spring 1996.

Rutchik, Gregory, *Japanese Research Projects and Intellectual Property Laws.* Office of Technology Policy, U.S. Department of Commerce, Washington, D.C., 1995.

Samuels, Richard, *Rich Nation, Strong Army: National Security and the Technological Transformation of Japan.* Cornell University Press, Ithaca, N.Y., 1994.

Sandholz, W., *High-Tech Europe: The Politics of International Cooperation.* University of California Press, Los Angeles, 1992.

Saxonhouse, Gary, "What Is All This about 'Industrial Targeting' in Japan?" *The World Economy,* vol. 6, 1983.

Saxenian, Annalee, *Regional Advantage: Culture and Competition in Silicon Valley and Route 128.* Harvard University Press, Cambridge, Mass., 1994.

Scalise, George M., *The Nature of High-Technology Competition,* 13 December 1995 contribution to Steering Committee deliberations.

Scherer, F.M., *International High-Technology Competition.* Harvard University Press, Cambridge, Mass., 1992.

Scott, Bruce R., *Economic Strategies of Nations,* in Charles W. Wessner (ed.), *Sources of International Friction and Cooperation in High-Technology Development and Trade, 30–31 May 1995, Papers and Proceedings.* National Academy Press, Washington, D.C., forthcoming.

Semiconductor Industry Association, *World Semiconductor Forecast,* Semiconductor Industry Association, San Jose, Calif., November 1995.

Servan-Schreiber, Jean-Jacques, *Le Defi Americain.* Edition de Noël, Paris, 1967.

Shear, Jeff, *The Keys to the Kingdom.* Doubleday, New York, 1994.

Sigurdson, J., *Industry and State Partnership in Japan: The Very Large Scale Integrated Circuits (VLSI) Project.* Discussion Paper No. 168, Lund, Sweden Research Policy Institute, 1986.

Smith, Hedrick, *Rethinking America.* Random House, New York, 1995.

Soete, Luc, "Technology Policy and the International Trading System: Where Do We Stand?" *Towards a New Global Framework for High-Technology Competition, 30–31 August 1995 Conference Proceedings.* Kiel Institute of World Economics, Kiel, Germany, forthcoming.

Solo, Robert A., "Gearing Military R&D to Economic Growth." *Harvard Business Review,* November–December 1962.

Spencer, Linda H., *Foreign Investment in the United States: Unencumbered Access,* Economic Strategy Institute, Washington, D.C., 1991.

Spencer, Linda M., *Foreign Acquisitions of U.S. High Technology Companies Database Report, October 1988—December 1993,* Economic Strategy Institute, Washington, D.C., March 1994.

Spencer, William J., *SEMATECH,* 20 November 1995 contribution to Steering Committee deliberations.

Spencer, William J., *Technology (Transfer) at SEMATECH,* 14 December 1995 contribution to Steering Committee deliberations.

William J. Spencer and Peter Grindley, "SEMATECH after Five Years: High-Technology Consortia and U.S. Competitiveness," *California Management Review,* vol. 35, Summer 1993.

Stalk, George, and Thomas M. Hoot, *Competing Against Time.* Free Press, New York, 1990.

Stern, John P., "Japan: the Philosophy of Government Support for Information Technology," in Charles Wessner (ed.), *Symposium on International Access to National Technology Programs, 19 January 1995 Symposium Proceedings.* National Academy Press, Washington, D.C., forthcoming.

Stern, Paula, "Reorganizing Government for Economic Growth and Efficiency," *Issues in Science and Technology,* Summer 1996.

Stern, Paula, *Getting the Boxes Right: New Blueprints for U.S. Economic Policymaking.* Economic Strategy Institute, Washington, D.C., 15 November 1995.

Stowsky, Jay, *Beating Our Plowshares into Double-Edged Swords: The Impact of Pentagon Policies on the Commercialization of Advanced Technologies.* Berkeley Roundtable on International Economics, Berkeley, Calif., 1986.

Stowsky, Jay, "Regional Histories and the Cycle of Industrial Innovation: A Review of Some Recent Literature." *Berkeley Planning Journal,* vol. 4, 1989.

Stowsky, Jay, and Burgess Laird, "Conversion to Competitiveness: Making the Most of the National Labs." *American Prospect,* Fall 1992.

Stremlau, John, "Dateline Bangalore: Third World Technopolis." *Foreign Policy,* Spring 1996.

Taoka, Shunji, (editorial) "The Whole Picture of the Domestic Patrol Aircraft Plan." *Tokyo AERA,* 12 February 1996.

Tassey, Gregory, *Technology and Economic Growth: Implications for Federal Policy.* NIST Planning Report 95-3, U.S. Department of Commerce, Washington, D.C., 1995.

Thurow, Lester, *Head to Head: The Coming Economic Battle among Japan, Europe, and America.* Morrow, New York, 1992.

Tilton, Mark, *Restrained Trade: Cartels in Japan's Basic Materials Industries.* Cornell University Press, Ithaca, N.Y., 1996.

Tonelson, Alan, "Beating Back Predatory Trade, *Foreign Affairs,* July–August 1994.

Tonelson, Alan, "The Perils of Techno-Globalism." *Issues in Science and Technology,* Summer 1995.

Trezise, Philip, "Industrial Policy is not the Major Reason for Japan's Success." *Brookings Review,* vol. 1, Spring 1983

Turner, Mark, "Labyrinth of Laws Could Lead to a Net Loss." *The Independent,* January 15, 1996.

Tyson, Laura, *Who's Bashing Whom? Trade Conflict in High-Technology Industries.* Institute for International Economics, Washington, D.C., 1992.

ul Haque, Irfan, *Trade, Technology, and International Competitiveness.* The World Bank, Washington, D.C., 1996.

United Nations Conference on Trade and Development, *Incentives and Foreign Direct Investment.* UNCTAD/DTCI/28, New York and Geneva, 1996.

U.S. Embassy, Tokyo, *International Market Insight Series,* "Research Budget," 960306

U.S. Technology Administration, *Key Foreign Industrial Competitors: Selecting Core-Technology Products.* Department of Commerce, Washington, D.C., October 1992.

United States Trade Representative, *1996 National Trade Estimate Report on Foreign Trade Barriers,* Washington, D.C., 1996.

Viner, Jacob, *Dumping: A Problem in International Trade.* University of Chicago Press, Chicago, 1923.

Wade, Robert, *Governing the Market: Economic Theory and the Role of Government in East Asian Industrialization.* Princeton University Press, Princeton, N.J., 1990.

Wade, Robert, "Managing Trade: Taiwan and South Korea as Challenges to Economic and Political Science," *Comparative Politics,* vol. 25, no. 2, January 1993.

Weiss, Linda, and John M. Hobson, *States and Economic Development: A Comparative Historical Analysis.* Polity Press, Oxford, U.K., 1995.

Wessner, Charles W. (ed.), *International Access to National Technology Programs, 19 January 1995 Symposium Proceedings.* National Academy Press, Washington, D.C., forthcoming.

Wessner, Charles W. (ed.), *Sources of International Friction and Cooperation in High-Technology Development and Trade, 30–31 May 1995, Papers and Proceedings.* National Academy Press, Washington, D.C., forthcoming.

White, Robert M., *U.S. Technology Policy: The Federal Government's Role.* The Competitiveness Policy Council, Washington, D.C., September 1995.

Willen, Paul, "Incomplete Markets and the Current Account Deficit." Economic Strategy Institute, Washington, D.C., 1996.

Williams, Scott, "Anti-Piracy Groups Target China—Technology Companies Want Federal Action." *Seattle Times,* 18 February 1994.

Wolff, Alan Wm., Thomas R. Howell, Brent L. Bartlett, and R. Michael Gadbaw (eds.), *Conflict among Nations: Trade Policies in the 1990s.* Westview Press, San Francisco, Calif., 1992.

The World Bank, *Bureaucrats in Business: The Economics and Politics of Government Ownership.* Oxford University Press, New York, 1995.

World Bank Policy Research Report, *The East Asian Economic Miracle: Economic Growth and Public Policy.* Oxford University Press, New York, 1993.

Wu, Yu-shan, *Comparative Economic Transformations: Mainland China, Hungary, The Soviet Union and Taiwan.* Stanford University Press, Stanford, Calif., 1994.

Ziegler, J. Nicholas, "A Capabilities-Based View of German Technology Policy." in Charles W. Wessner (ed.), *Sources of International Friction and Cooperation in High-Technology Development and Trade, 30–31 May 1995, Conference Proceedings.* National Academy Press, Washington, D.C., forthcoming.

Zysman, John, *Political Strategies for Industrial Order: State, Market and Industry in France.* University of California Press, Berkeley, 1977.

V. National Technology Policies and International Friction: Theory, Evidence, and Policy Options[*]

This section of the Report addresses the theory and empirical evidence of national technology policy in globalized markets, and it deals with approaches to multilateral conflict resolution in selected policy areas giving rise to international friction: subsidies and public procurement, market access and structural impediments, and dumping, antidumping, and competition policies. Prepared as an analytical basis for deliberations of the Steering Committee, it is intended to give a theoretical and empirical underpinning to the Committee's Recommendations and Findings.

Economic theory and empirical evidence suggest that governments could usefully intervene in high-technology in two ways. First, they could act as a neutral agent that creates the necessary credibility, commitment, and mutual trust among private companies to facilitate their cooperation in high-risk, high-volume R&D. Second, if in view of the externalities involved an element of subsidization is to be added, this should be done in a nondiscriminatory fashion. A favorable tax treatment of R&D may be the most appropriate tool to achieve this task. In practice, whether justified on economic grounds or not, governments *do* engage in targeted industrial and technology policies. This gives rise to recurrent trade friction.

To resolve that friction this section of the Report suggests, among other measures, an open subsidies club where access to the public funding of

*Section V, copyright 1996 by the Hamburg Institute for Economic Research and the Kiel Institute for World Economics. All rights reserved.

high-technology development should be open to domestic as well as foreign firms on a reciprocal basis. To overcome structural impediments to high-technology trade and investment, the multilateral route offered by the so-called nonviolation clause of Article 23 GATT should be given preference over unilateral approaches. In order to reduce the potential for protectionist misuse of the antidumping instrument, proceedings should make explicit reference to the state of competition in the relevant exporting and importing country markets, thus introducing elements of competition policy into the proceedings. To meet specific antitrust concerns in high-technology, relating inter alia to network externalities, systems leverage, standardization and innovation cartels, the paper supports the *Draft International Antitrust Code (DIAC)* proposed by an international group of legal experts.

The section was jointly produced by staff members of the Kiel Institute of World Economics (IfW) and the HWWA Institute for Economic Research. The first chapter, *The Economics of Technology Policy in Globalized Markets,* and the first part of the second chapter, *Approaches to Conflict Resolution: Agenda for Action,* were written by Karl-Heinz Paqué, Jürgen Stehn, and Ernst-Jürgen Horn of IfW. The latter two sections of chapter 2 were written by Hans-Eckart Scharrer and Georg Koopmann of HWWA. The second section draws on the papers and discussions at the three conferences held in Hamburg, Washington, D.C., and Kiel, and it greatly benefited from the discussions of the Steering Committee. The authors were also primarily responsible for the scientific programs of the two German conferences and the editing of the proceedings, which are published as separate volumes.* Their collaborative effort, in partnership with their counterparts at the NRC, has been critical to the success of this joint venture in transatlantic research cooperation.

Erhard Kantzenbach
Project Co-Chairman

---

*See George Koopman and Hans-Eckart Scharrer (eds.), *The Economics of High-Technology Competition and Cooperation in Global Markets.* Baden-Baden, 1996, and Horst Siebert (ed.), *Towards a New Global Framework for High-Technology Competition.* Tübingen, forthcoming.

V. JOINT HWWA-IFW ANALYSIS: NATIONAL TECHNOLOGY POLICIES AND INTERNATIONAL FRICTION: THEORY, EVIDENCE, AND POLICY OPTIONS

The Economics of Technology Policies
in Globalized Markets

ANALYTICAL BENCHMARKS

To a large extent, international frictions in markets for high-technology products are the consequence of some form of unilateral state intervention. Using a large variety of instruments, many governments grant support to specific branches of economic activity whose growth is deemed to be beneficial or even crucial for the long-term performance of the national economy. The policy toolbox ranges from overt protectionism and subsidization to more covert barriers to market entry such as discrimination in public procurement, product standards, and distributional networks. Whatever their specific shape may be, the instruments are usually part of a more or less coherent growth strategy that is well summed up under the terms "technology policy" or "industrial policy."[1]

Many of the more popular arguments for technology policies in high-technology markets can easily be refuted with standard economic reasoning and are therefore no serious candidates for further scrutiny.[2] The more powerful case for technology policy—and the one that is most likely to dominate future debate on trade frictions in high-technology markets—comes from two modern strands of economic research that are both analytically important and politically relevant: endogenous growth theory and strategic trade theory.

Endogenous Growth Theory

Some strands of endogenous growth theory—notably the pioneering work by Grossman and Helpman (1991) and by Rivera-Batiz and Romer (1991a, 1991b)[3]—have focused on the link between international trade integration

---

[1] We use these terms interchangeably in the following paragraphs, with both terms describing a government policy that grants support to a (specific) high-technology industry.

[2] A standard undergraduate textbook on international trade—P. Krugman, and M. Obstfeldt, *International Economics: Theory and Policy,* 3rd ed. New York, 1994, pp. 278–281—does an excellent job in refuting some common errors and misapprehensions.

[3] G.M. Grossman and E. Helpman, *Innovation and Growth in the Global Economy,* Cambridge, Mass., 1991; F.L. Rivera-Batiz, P.M. Romer, "Economic Integration and Endogenous Growth," *Quarterly Journal of Economics,* vol. 106 (1991) pp. 531–556, and "International Trade with Endogenous Technological Change," *European Economic Review,* vol. 35 (1991), pp. 971–1004. For a summary evaluation of the policy relevance of the new growth theories, see also F.L. Rivera-Batiz, *The Economics of Technological Progress and Endogenous Growth in Open Economies.* Paper presented at the conference *The Economics of High-Technology*

and the concomitant changes of sectoral specialization patterns on one side and the long-run growth prospects of a country on the other. This work contains a vast array of rather complex analytical insights that may also be of interest for selected questions of economic policy. With some coura-geous simplifications, the policy gist of the work—as far as it is relevant for technology policy in advanced economies—may be summarized as follows.[4]

International trade patterns in (free) high-technology markets are deter-mined by comparative advantages. More than in other markets, however, these comparative advantages may be affected by the resources that differ-ent economies devote to industrial research, i.e., to investment in the cre-ation of new knowledge. But not each and every piece of new industrial knowledge affects trade patterns; whether it does depends, most impor-tantly, on the subsequent diffusion of this knowledge, i.e., in economists' jargon, on the degree of localization of technological spillovers. In this respect, two polar cases deserve attention as conceptual benchmarks: knowledge as a global good and knowledge as a national good.

If the spillovers are essentially global, i.e., if competitors in all relevant countries have access to any addition to the (common) knowledge pool, wherever it comes from, there will be no lasting effects of national research efforts on trade patterns, and one is back at standard explanations of trade in terms of national endowments, with ordinary production factors, notably labor, human capital, and physical capital. In these circumstances, national technology policy makes little economic sense because what it helps to create in terms of new knowledge will easily diffuse outside the country, without giving domestic producers a viable and persistent advantage over foreign competitors.

If, on the other hand, knowledge spillovers remain geographically con-centrated and thus essentially national in scope, a cumulative process of what may be called "national learning" may then set in and drive a widen-ing competitive wedge between the respective national industry and the rest of the world. In the extreme, a country starting with a tiny and accidental technological lead may eventually dominate the relevant world market be-cause it profits—alone and persistently—from its own knowledge creation,

---

Competition and Corporation in Global Markets held at the HWWA-Institute in Hamburg, 2–3 February 1995; and R. Baldwin, *The Economics of Technological Progress and Endogenous Growth in Open Economies.* Comment presented at the conference "The Economics of High-Technology Competition and Cooperation in Global Markets," HWWA-Institute in Hamburg, 2–3 February 1995; L. Soete, *Technology Policy and the International Trading System: Where Do We Stand?* Paper presented at the conference "Towards a New Global Framework for High-Technology Competition," in Kiel, 30–31 August 1995.

[4] A detailed survey of results is given by M. Stolpe, *Technology and the Dynamics of Specialization in Open Economies*, Kiel Studies 271, Tübingen, 1995, Section B. IV, pp. 41–47.

which allows it to lower costs, to raise quality levels, and to introduce new products and production processes. In these circumstances, a national technology policy may well have a powerful rationale: if private producers do not make the socially optimal decisions—in economists' jargon, if there is a market failure—government may step in to initiate the virtuous high-technology growth circle.

Is the market likely to fail? Among the cases of potential market failure that have been identified in the literature, two stand out in importance, one focusing on positive externalities of private R&D spending, the other on negative ones. The argument on positive externalities recognizes a tendency toward private *under*investment in R&D due to any positive spillover contributions to the (national) stock of knowledge that cannot be appropriated in a private calculus of profit maximization. The government is then called upon to raise the level of research effort in the industry at hand.

The one on negative externalities identifies a tendency toward *over*investment due to inefficient parallel research: with a limited common pool of potential discoveries and innovations at any point in time, a successful innovation is likely to reduce the prospective commercial value of research efforts by other firms, a form of negative (pecuniary) externality that is not properly taken into account by private agents.[5] The government may then be called upon to reduce, or better, to bundle and focus research efforts so as to ensure a maximum expected social rate of return and a minimum deadweight loss.

Which instruments of intervention should the government use? In the case of positive externalities, the modern theories of endogenous growth recommend public support directly of R&D activities rather than general protectionist measures for the respective industrial branch, say, in the form of production subsidies or tariffs. This is in line with the more traditional welfare analysis of externalities in trade theory,[6] though the underlying rationale of the result is somewhat different: while traditional theory wants the government to avoid static allocative losses, the modern theory wants it to avoid a (growth-hindering) resource competition for skilled labor that may be used both in manufacturing and in research. E.g., production subsidies or tariff protection for the production of a high-technology good may induce highly qualified engineering personnel to move from research into production, thus increasing the cost of R&D and reducing the country's potential for growth.

---

[5] This (pecuniary) externality is a typical result of models of (Schumpeterian) creative destruction as pioneered by P. Aghion and P. Howitt, "A Model of Growth through Creative Destruction," *Econometrica*, vol. 60 (1992), pp. 323–351.

[6] See, e.g., the classical statement by M. Cordon, *Trade Policy and Economic Welfare*, Oxford, 1974, pp. 257–264.

A similar argument for direct intervention applies in the case of negative externalities, although it has not yet been spelled out in detail in the literature: if a "bundling" of R&D efforts is required to reduce inefficient parallel research, it should be done by allowing firms to cooperate so as to rationalize and coordinate some of their research investment, possibly under government auspices and encouragement.

Strategic Trade Theory

Some modern strands of trade theory—beginning with the pioneering work by Brander and Spencer (1983, 1985) and Dixit and Kyle (1985)[7] — have focused on the international rivalry for monopoly rents in world markets that operate under conditions of highly imperfect competition, usually involving only very few producers from different countries. This rivalry can have the characteristic that possibly accidental initial advantages of one firm lead to high monopoly profits because potential competitors are deterred from market entry by high start-up costs and/or the narrowness of the prospective market.

Once again, the details of the relevant theoretical models are complex, but the case for government intervention in the form of so-called strategic trade policy is simple and straightforward: by granting a temporary subsidy to a newcomer, the government may turn potential competitors into actual ones, thus breaking up a (quasi-)monopolistic market position of the dominant foreign producer(s) and shifting at least some of the monopoly rents from one country to the other. If, in the end, national subsidy costs are lower than the gain in rents, then the active policy stance pays off from a national point of view.

Note that, in the technology policy debate, the case for a strategic trade policy hardly comes neatly separated from arguments based on endogenous growth theory. E.g., the case for the European Union's long-standing effort to break up the American quasi monopoly in the world market for commercial aircraft has been consistently justified on two grounds: shifting rents across the Atlantic *and* laying the ground for the expansion of a high-technology industry that was regarded as greatly beneficial in terms of internal and external learning processes for future economic growth in Europe. In fact, the American lead in aircraft and related technologies was widely identified as being the consequence of a costly learning process that any prospective European competitor would still have to go through.

---

[7] J.A. Brander, B. J. Spencer, "International R&D Rivalry and Industrial Strategy," *Review of Economic Studies*, vol. 50 (1983), pp. 707–722, and "Export Subsidies and International Market Share Rivalry," *Journal of International Economics*, vol. 16 (1985), pp. 83–100; A.K. Dixit and A.S. Kyle, "The Use of Protection and Subsidies for Entry Promotion and Deterrence," *American Economic Review*, vol. 75 (1985), pp. 139–152.

EMPIRICAL EVIDENCE

What do we know about the prospects for a successful technology policy in actual reality? Or more specifically: what do we know about the extent to which there are cumulative processes of external learning that are geographically concentrated? And what do we know about the extent to which there are monopoly rents to be shifted around internationally? It may not come as a surprise that any serious economist can hardly avoid answering all of these questions with "very little." Our current ignorance simply reflects the enormous problems of empirically isolating and measuring the relevant phenomena, not to speak of identifying causal relationships.

To be sure, there is a large literature of individual case studies, in which selected high-technology branches or markets are picked to demonstrate the success or failure of particular government measures. In the context of the bilateral trade disputes between the United States and Japan, some of these studies have received considerable public attention, notably the book by Tyson (1992).[8] These studies suffer from major methodological deficiencies: despite their wealth of information and interpretation, they offer neither a theory-based empirical account of the diffusion of knowledge or the shifting of rents, nor a reasonably specified counterfactual scenario that would allow pinning down with some conceptual precision to what extent government intervention has in fact altered the path of economic history.

The example of Japanese industrial policies is notorious in this respect.[9] To qualify as a valuable piece of economic analysis, it is *not* sufficient to demonstrate that there was some intervention in some industrial branch and that producers in this branch were increasingly successful in world markets; rather, it has to be shown that the relevant path of events was sufficiently different from what could have been expected on the basis of factor endowments and learning processes that went on anyway, independently of government action.

It is an a priori open question whether the remarkable world market penetration that Japanese industry achieved in the mass production of cameras, automobiles, and semiconductors was the "natural crop" of the specific engineering skills that the Japanese education and training system tends to provide (like, say, the German or the Swiss one) or whether it was the

---

[8] L.D. Tyson, *Who's Bashing Whom? Trade Conflict in High-Technology Industries.* Washington, D.C., 1992. See also, most recently, S. Cohen and P.H. Chin, *Tipping the Balance: Trade Conflicts and the Necessity of Managed Competition in Strategic Industries.* Paper presented at the conference "Towards a New Global Framework for High-Technology Competetion, Kiel, 30–31August 1995.

[9] On Japanese industrial policies in the relevant periods, see K. Yamamura, "Caveat Emptor: The Industrial Policy of Japan," in P. Krugman (ed.), *Strategic Trade Policy and the New International Economics*, Cambridge, Mass., 1986.

"artificial consequence" of smart technology policy initiatives—or both. To come closer to an answer, one would have to study carefully the dynamics of comparative advantages over time and across countries, testing for elements of path dependence and hysteresis in specialization patterns of production and trade,[10] and linking this to government interventions. Clearly, the case studies available so far fall well short of this standard.

Apart from the more descriptive case studies, however, there is a small but growing literature that makes a serious econometric attempt at identifying phenomena that are relevant for technology policymaking. In the following paragraphs, we will provide a brief review of important recent pieces of empirical analysis—first of those concerning external learning effects and growth, and second of those concerning rent-shifting through strategic trade policy.[11]

External Learning and Growth

There has been a large number of studies over the last three decades which, in one way or another, try to measure and quantify spillover effects of R&D. While many of them are flawed and subject to a variety of reservations, they do on the whole support the view that R&D spillovers are positive and quantitatively important.[12] We shall focus strictly on those (relatively few and recent) studies that investigate the regional and sectoral incidence of learning effects and externalities, which is the core issue for the design of technology policies.

The Case of the Semiconductor Industry

In recent years, the semiconductor industry has been at center stage of the technology policy debate, not least because it is widely regarded as a strategic industry in the sense that the production of semiconductors involves strong learning effects (and thus cost reductions) over time, and that semiconductors are used as high-technology inputs in virtually all branches

---

[10] An econometric step in this direction has been made by M. Stolpe, *Technology and the Dynamics of Specialization in Open Economies*, Kiel Studies 271, Tübingen, 1995, chaps. 3 and 4.

[11] For a more extensive survey, see K.H. Paqué, "Technologie, Wissen und Wirtschaftspolitik— Zur Rolle des Staates in Theorien des endogenen Wachstums," *Die Weltwirtschaft* 3 (1995), pp. 237–253.

[12] This is also the conclusion of the most comprehensive survey of this literature to date: Zvi Griliches, "The Search for R&D Spillovers," *Scandinavian Journal of Economics*, vol. 94 (1992), Supplement, pp. 24–47.

of any modern economy.[13]   Irwin and Klenow (1994a)[14] deliver the first serious econometric attempt to estimate the actual magnitude of internal and external learning effects as well as the geographic concentration of externalities for the semiconductor industry.

On the basis of a large set of quarterly data for the period 1974–1992 and for roughly thirty semiconductor producers in the United States, Europe, and East Asia (Japan and South Korea), the study provides estimates of how strongly the product price depends on past production of semiconductors (1) in the firm itself, (2) in the country where the firm is located, and (3) in the world as a whole. In the absence of cost data, the product price is taken as a proxy for "dynamic" marginal production cost,[15] i.e., the marginal cost that takes into account the discounted value of all future cost reductions due to production experience. Cumulative past production is taken as a proxy for the cumulative production experience and thus for the level of technical knowledge, i.e., the stage of the learning process. Also, an explicit distinction is made between cumulative experience within a production line—i.e., within the current generation of memory chips—and the experience with earlier production lines, i.e., with older memory chip generations.

The results of Irwin and Klenow (1994a) are remarkable and important. They find strong learning effects *within* each chip generation—on average a 20 percent cost reduction, with a doubling of output over time. However, they find no significant learning effects between chip generations: any new quality stage of technological development in the form of a new generation of memory chips begins with a level playing field. They also find rather powerful externalities: additional output (and thus additional experience) of other firms leads to a cost reduction of roughly one-third of a corresponding output increase of the firm itself. What they do not find, however, is a significant difference between *inter*national and *intra*national externalities: knowledge spillovers—as far as they exist—appear to be undisturbed by national borders. Also, the authors of the study do *not* find any significant

---

[13] See T. F. Bresnahan and M. Trajtenberg, *General Purpose Technologies: "Engines of Growth"?* NBER Working Paper No. 4148, National Bureau of Economic Research, Cambridge, Mass., August 1992; and E. Helpman and M. Trajtenberg, *A Time to Sow and a Time to Reap: Growth Based on General Purpose Technologies.* NBER Working Paper No. 4854, National Bureau of Economic Research, Cambridge, Mass., September 1994, who provide a rigorous analysis of the link between economic growth and the expansion of what they call "general purpose technologies," i.e., technologies that are "extremely pervasive" in the sense of being used as inputs by a wide range of sectors of an economy.

[14] D.A. Irwin and P.J. Klenow, "Learning-by-Doing: Spillovers in the Semiconductor Industry," *Journal of Political Economy*, vol. 102 (1994a), pp. 1200–1227.

[15] Strictly speaking, it is an "adjusted" product price that is used: Irwin and Klenow, op. cit., pp. 1212–1213, assume Cournot-competition and adjust the price accordingly, thus taking account of changing monopolistic mark-ups over the product cycle.

difference in the structural parameters of learning and of external effects between Japanese and other firms.

All in all, these results cast serious doubt on whether the popular idea that there are powerful first-mover advantages for early market entrants is really compatible with the empirical record of the semiconductor industry. Prima facie, the simple observation that Japanese semiconductor producers—after a prolonged period of spectacular world market penetration—lost world market shares back to American competitors by the late 1980s, seems to point in the same direction. However, the interpretation of this fact is complicated by two major policy shifts that may have influenced the course of events, namely the conclusion of two U.S.-Japanese semiconductor trade agreements, which brought some protectionist relief to American producers, and the establishing of SEMATECH, the joint industry-government research consortium in the American semiconductor industry.

With respect to the trade agreements, there is by now a broad consensus that they came too late—the first in 1986—to prevent a fundamental restructuring of American industry away from the mass production of DRAMs[16] to more profitable market segments (e.g., design-intensive chips). With the benefit of hindsight, one can say that this turned out to be a blessing because of the rising tide of competitive pressures from South Korean producers that forcefully entered the DRAM market and undermined the dominating Japanese position.[17]

With respect to the economic impact of SEMATECH, there has been a first econometric attempt at estimation, again by Irwin and Klenow (1994b).[18] Using data on American semiconductor firms, i.e., on those which were members of SEMATECH in the relevant period and those which were not (and which, statistically, form a control group), the authors estimate the effects of the consortium on the members' R&D spending, profitability, investment, and productivity. Their results turn out to be conclusive only for the effect on R&D spending: they indicate that SEMATECH has *reduced* R&D spending by roughly 300 million U.S. dollars per year. In terms of the two alternative theoretical interpretations of technology policy presented above,[19] this supports the view that SEMATECH has served as an

---

[16] Dynamic Random Access Memory chips.

[17] On the genesis and the details of the American-Japanese semiconductor agreement, see D.A. Irwin, *Trade Politics and the Semiconductor Industry*. NBER Working Paper No. 4745. National Bureau of Economic Research, Cambridge, Mass., May 1994, pp. 69–71; for a summary of opinions, see I. Maitland, "Who Won the Industrial Policy Debate?" *Business & the Contemporary World* 1 (1995) pp. 83–95, notably pp. 88–89.

[18] D.A. Irwin, and P.J. Klenow, *High Tech R&D Subsidies: Estimating the Effects of SEMATECH*. NBER Working Paper No. 4974. National Bureau of Economic Research, Cambridge, Mass., December 1994(b).

[19] See the section "Endogenous Growth Theory" above.

instrument to focus and bundle research, i.e., to reduce inefficient parallel efforts, rather than to expand the research scope into ranges that would otherwise not have been profitable for private member firms to explore.[20]

While SEMATECH may thus have helped firms to coordinate their research efforts efficiently, there is by now broad agreement that this piece of positive technology policy can hardly be made responsible for the bulk of the recovery of the American semiconductor industry.[21] Instead, the timely restructuring of the industry appears to have done by itself the major part of the work. In view of the econometric evidence about the powerful international diffusion of technological knowledge in this high-technology industry, one should not be surprised to see one country's producers recover quite successfully even from rather deep adjustment crises.[22]

Knowledge Diffusion: International and Interregional

Coe and Helpman (1993)[23] attempt to quantify international knowledge spillovers between industrial economies.[24] On the basis of annual data for the period 1971–1990 and twenty-two mostly-OECD countries, they estimate how strongly total factor productivity in any country depends on the "research capital stock" (1) in this country itself and (2) in all other countries. The domestic "research capital stock" is defined as total national R&D expenditure, accumulated over time using a certain rate of "knowledge depreciation;" the total foreign research capital is calculated as a weighted average of that of all other countries, with the weights being bilateral import quotas, i.e.—roughly speaking—the degree of trade integration.

---

[20] Irwin and Klenow (1994b) speak of support for the "sharing hypothesis" as against the "commitment hypothesis."

[21] See, inter alia, the assessment in *The Economist*, 2 April 1994, "Uncle Sam's Helping Hand," pp. 91–93

[22] It is most unfortunate that no methodologically comparable studies are available for evaluating specific technology policies of other countries, notably those of Japan and the European Union. This is so because, to our knowledge, no data are available that would allow the setting up of control groups of firms that are not covered by the relevant policy and the quantifying of the effect of government intervention against this background. Thus the many available accounts of European technology policy—e.g., H. Klodt, *Wettlauf um die Zukunft*, Kieler Studien No. 206, Tübingen, 1988; H. Klodt et al., *Forschungspolitik unter EG-Kontrolle*. Kieler Studien No. 220, Tübingen, 1988; and, more recently, H. Klodt and J. Stehn et al., *Strukturpolitik der EG*, Kieler Studien No. 249, Tübingen, 1992, pp. 98–114 and 152–160— have to remain largely descriptive and interpretative.

[23] D.T. Coe, and E. Helpman, *International R&D Spillovers*, NBER Working Paper No. 4444, National Bureau of Economic Research, Cambridge, Mass., August 1993.

[24] A methodologically comparable study that focuses, however, on spillovers from rich to poor countries is D.T. Coe, E. Helpman, and A.W. Hoffmaister, *North-South R&D Spillovers*, CEPR Discussion Paper No. 1133, Centre for Economic Policy Research, London, February 1995.

The authors find that total factor productivity in a country depends strongly on the domestic research capital: according to their estimates, a one percent increase of the research capital stock leads to a 0.25 percent productivity increase. They also find that foreign research capital matters as well, but that it matters comparatively more for smaller countries than for large ones. In fact, for small highly integrated economies, research efforts in the main trading-partner countries appear to matter at least as much as corresponding efforts at home. Apparently, trade between advanced economies—and with it international capital movements—appear to be powerful channels for transmitting technological knowledge across borders, a result which tends to support the estimates and message of Irwin and Klenow (1994a) in the narrower context of the semiconductor industry.

On interregional knowledge spillovers, there are two major studies that are directly relevant to the policy questions at hand. Jaffe, Trajtenberg, and Henderson (1993)[25] exploit the geographic information contained in U.S. patent statistics to draw econometric inferences on the extent to which the diffusion of knowledge remains localized after an innovation has been made in some part of the country. The central piece of information on which they build their econometrics is the geographic pattern of patent citations to be found in new patent applications, which can serve as a kind of roadmap to track knowledge spillovers.

While the methodology is rich and complex in detail,[26] the results are rather clearcut. First of all, they point toward quite strong localization effects, with the degree of localization being higher for patents of private firms than for those of universities, which is plausible because purely academic research results are likely to be circulated more openly than the knowledge created in private research laboratories. Second, for any patent, localization tends to decline over time, but very slowly: even a decade after a patent is granted, the geographic diffusion pattern remains little changed for all practical purposes. And third, the high degree of *geographic* localization is *not* matched by a corresponding degree of *sectoral* localization: whatever the grouping of patents into technological or industrial segments, there is always a remarkably high share of citations that refer to patents in very different fields. This casts some doubt on the common assumption that cumulative learning processes *within* a well-defined high-technology industry are a valid rationale for technology policy.

---

[25] A.B. Jaffe, M. Trajtenberg, and R. Henderson, "Geographic Localization of Knowledge Spillovers as Evidenced by Patent Citations," *Quarterly Journal of Economics*, vol. 108 (1993), pp. 577–598.

[26] See Jaffe, Trajtenberg, and Henderson, op. cit., pp. 580–585. The main methodological problem is to separate genuine spillovers from correlations that are due to a preexisting pattern of geographic concentration of technology-relevant activities.

This last result has received independent support from another strand of research on interregional knowledge spillovers, which applies ideas of endogenous growth theory to the study of urban agglomerations. In a major contribution on the economic determinants of city growth, Glaeser et al. (1992)[27] test two competing hypotheses, one identifying knowledge spillovers *within* an industry, the other one spillovers *between* industries as the driving force of output and employment growth. Using a large data set on the structure of industry for 170 metropolitan areas in the United States for the period 1956–1987, the authors estimate to what extent the long-term growth of a city industry was positively or negatively correlated with a number of characteristics of local industry structure, most importantly the degree of specialization (1) of the respective industry and (2) of the rest of the urban economy.

The results are again rather clearcut: other things being equal, an industry tends to grow faster in cities where it is still underrepresented and where the rest of the urban economy has a low degree of industrial specialization. Hence, as in the research on patent knowledge diffusion, an empirical case can be made for inter-industrial rather than intra-industrial spillovers dominating the picture.[28]

Strategic Trade Policy: The Aircraft Industry

Rent-shifting in high-technology markets through deliberate government intervention has been an explicit aim of policy in one particular branch of economic activity: the aircraft industry. Naturally, it has been this industry and in particular the economics of the European launching of Airbus as a competitor of Boeing that was subjected to empirical analysis, notably by

---

[27] E.L. Glaeser, H.D. Kallal, A. Scheinkman, and A. Shleifer, "Growth in Cities," *Journal of Political Economy*, vol. 100 (1992), pp. 1126–1152.

[28] Note, however, that—by not specifying any proxy variable for the state of "knowledge"—the methodology used by Glaser et al. (1992) may be subject to the criticism that it does not allow to discriminate between technological externalities and mere pecuniary externalities, which work through the price mechanism and do less easily qualify as a rationale for technology policy. The same problem plagues another strand of research which tries to identify intersectoral external effects by examining the input-output relationships between industries and estimating to what extent the productivity in one industry is affected by output changes in other industries. By this token, R.J. Caballero, and R.K. Lyons, "The Case for External Effects," in A. Cukierman, (ed.), *Political Economy, Growth, and Business Cycles*, Cambridge, Mass., 1992, pp. 117–139, find sizable externalities, but it remains unclear of what kind. On these methodological problems in detail, see M. Stolpe, *Technology and the Dynamics of Specialization in Open Economies*, Kiel Studies 271, Tübingen, 1995, Section B. V, pp. 51–54.

Baldwin and Krugman (1988) and Klepper (1990, 1994).[29] For a number of reasons ranging from the duopolistic market structure to simple lack of data, all these empirical studies consist of model calibration and simulation rather than econometric estimation of theory-based parameters.[30] This is why all results must be interpreted with utmost caution: because they depend crucially on the models' assumptions and imputed parameters.

The policy-relevant core of these studies consists in answering the following double question: to what extent and for what economic reason did the subsidized market entry of Airbus—and thus the transformation of the aircraft market from a (prospective) monopoly of Boeing into a transatlantic duopoly—lead to a redistribution of producer and consumer surplus in the United States, in Europe, and in the rest of the world? Despite considerable methodological differences, the answers to these questions do not differ greatly between the relevant studies.

Briefly summarized, the answers read as follows. The market entry and continued market presence of Airbus has led or will lead to a loss of producer surplus in the United States due to the reduction of (monopoly) profits and a gain of consumer surplus all over the world due to lower aircraft prices. However, European "consumers/taxpayers" are likely to end up worse off because the additional tax burden may well overcompensate for the gain in consumer surplus. Also, world welfare as a whole is likely to be reduced because the loss of producer surplus in the United States may well overcompensate for the worldwide gains in consumer surplus (net of the tax burden), a result which reflects the enormous importance of economies of scale in the aircraft industry: given very sharp cost reductions through learning effects, the socially optimal outcome for the world as a whole may simply be a monopoly.

The central conclusion from these results is that, in economic terms, the European Airbus venture is better interpreted as a worldwide anti-monopoly policy than as a transatlantic rent-shifting: European consumers/taxpayers foot the bill of breaking an American monopoly to the unambiguous advan-

---

[29] R. Baldwin, P.R. Krugman, "Industrial Policy and International Competition in Wide-Bodied Jet Aircraft," in R. Baldwin (ed.), Trade Policy Issues and Empirical Analysis, Chicago, 1988; G. Klepper, "Entry into the Market for Large Transport Aircraft," *European Economic Review*, vol. 34 (1990), pp. 775–803; "Industrial Policy in the Transport Aircraft Industry," in P. Krugman, and A. Smith (eds.), *Empirical Studies of Strategic Trade Policy*, Chicago, London, 1994, pp. 101–126. A methodologically comparable analysis of the market for small-size aircraft is provided by R. Baldwin, and H. Flam, "Strategic Trade Policies in the Market for 30–40 Seat Commuter Aircraft," *Weltwirtschaftliches Archiv/Review of World Economics*, vol. 125 (1989), pp. 484–500.

[30] On the specific methodological problems in this field of empirical research, see P. Krugman, "Introduction," P. Krugman and A. Smith, op. cit., pp. 4–5.

tage of consumers in the rest of the world.[31] Whether one likes this policy or not, it has very little to do with the idea of a genuine strategic trade policy that postulates a national interest in rent-shifting as the normative basis for government intervention, and not an altruistic policy stance vis-à-vis the rest of the world.

Needless to say, one may put forward other arguments for fostering a European aircraft industry as an important high-technology branch, but these usually lead into the realm of technological spillovers (see "External Learning and Growth" above). To our knowledge, there have been no systematic empirical assessments of the aircraft industry with respect to technological externalities. Casual observations—e.g., the apparent failure of Germany's Daimler-Benz to profit from so-called synergy effects between aircraft and motor car production[32]—suggest that these effects have been grossly overestimated in the past and have lured some firms into a path of diversification that turned out to be unprofitable in the longer run.

There is a more general conclusion to be drawn from the empirical studies on rent-shifting that goes well beyond the aircraft market. Obviously, the aircraft industry is almost a textbook case for an industry with strong economies of scale and, consequently, very few commercial players: and fat monopoly margins: if even in this industry it is very hard to channel rents into the pockets of producers in the intervening country, how can one ever arrive at a powerful case for government intervention on these grounds in other industries that are much further away from the conditions of a natural monopoly? One may suspect that the chances to do so are small at best.

TECHNOLOGY POLICY:
SOME CAUTIOUS CONCLUSIONS

Despite the vast uncharted territories that still await future research—notably in the field of econometrically scrutinizing the various dimensions of knowledge diffusion—some preliminary conclusions may be drawn from the available evidence surveyed above.

---

[31] Note, however, that this result holds only if Boeing would in fact have become a monopoly without the subsidized market entry of Airbus. Yet this is an open question because Boeing's intra-American competitor McDonnell Douglas might have remained in the market in this case. A model calibration by Klepper (1990, 1994), which explicitly puts the actual Airbus entry against a duopoly with equal market shares of Boeing and McDonnell Douglas, yields the counterintuitive result that a duopoly Boeing/Airbus leaves Boeing with *higher* profits than the duopoly Boeing/McDonnell Douglas because the latter involves more equal market shares and thus forces Boeing further up the average cost curve. Whatever one may think of this peculiar result—it depends crucially on the imputed market share of McDonnell Douglas, which is purely speculative—the Airbus venture loses its antimonopoly rationale once it is assumed that there is no monopoly in the counterfactual scenario anyway.

[32] See, e.g., *The Economist,* 20 May 1995, "Schrempp Cocktail" (p. 70), and 18 November 1995, "A Tale of Two Conglomerates" (p. 20) and "Dismantling Daimler" (p. 79).

(1)   Knowledge spillovers appear to be a pervasive feature of modern economies that is likely to be of great importance for economic growth. There is still much less clarity about how the knowledge diffusion actually works, although there are some patches of relevant evidence: studies on the inter*regional* diffusion process point to strong localization effects in terms of geography, but not in terms of sectors of economic activity; in turn, studies on inter*national* diffusion point to powerful spillovers across national borders, be it on a macroeconomic level or for a selected important high-technology industry such as semiconductors. The apparent contradiction between the empirical results from interregional and international studies may have its cause in a genuine "globalization" of knowledge flows: it is not implausible to suspect that the leading high-technology centers in different countries may be better linked in terms of communication than the high-technology centers and the periphery in one single (large) country.

(2)   The aim of shifting monopoly rents between countries does not by itself make a case for government intervention. The one instance where it was obviously relevant—the aircraft industry—is history, since Airbus is by now an established competitor of the former (quasi-) monopolist Boeing. Whether the launching of Airbus was a policy success depends on the criteria used. Be that as it may, no comparably structured high-technology branch, where fat monopoly profits accrue in just one country, is in sight in the near future.

(3)   The actual record of government intervention in high-technology industries is very hard to evaluate because, usually, no sensible counterfactual scenario can be made available. However, there is one important case where such a scenario has been at least tentatively constructed: SEMATECH. The relevant analysis of SEMATECH indicates that the consortium helped to reduce—arguably inefficient—parallel research, thus pointing to a positive role for the government as an agent that bundles and focuses rather than expands research efforts.

Back to the basic question: is there a theoretically sound and empirically supported rationale for government intervention in high-technology markets? Given the extent of our ignorance of the precise determinants and structure of knowledge flows, it is hard to avoid the conclusion that a government in an advanced economy has exactly the same information problem as the empirical economist in tracing the relevant knowledge flows. Hence a fine-structured industrial policy targeted at selected high-technology industries can hardly be recommended.[33] On the other hand, the observed regional localization of knowledge spillovers may suggest that a "technol-

---

[33]   This conclusion is almost a commonplace among economists. See, e.g., the policy conclusions by one of the main advocates of endogenous growth theory: G.M. Grossman, "Promoting New Industrial Activities: A Survey of Recent Arguments and Evidence," OECD Economic Studies, vol. 14 (1990), pp. 117–119.

ogy policy" aimed at generally supporting the (otherwise suboptimal) growth of high-technology agglomerations, can make economic sense. The problem is, of course, how this should be done with a minimum risk of incurring deadweight losses for taxpayers and consumers. Some reasonable guidelines may read as follows.

(1)   In selecting instruments of government intervention, one should aim at supporting R&D itself, and not output or trade, because output subsidies and trade policies invariably have undesirable allocative side effects, not to speak of the potential for political frictions that may come to the fore once countries engage in protectionist warfare in high-technology markets.

(2)   When choosing R&D for government support, one should consider whether the apparent market failure cannot be corrected through what may be called "government coordination" rather than through subsidies. Finding an economically efficient level and structure of industry research by jointly launching particularly large and risky projects or by weeding out costly parallel research is eventually in the interest of all firms concerned; and at least in high-technology markets that are characterized by (not too wide) oligopolies, firms may be ready to cooperate in research. Hence the government may simply serve as a positively neutral agent that creates the necessary credibility, commitment, and mutual trust among the private parties to make the joint venture possible at all. With the benefit of hindsight, SEMATECH may in fact be interpreted in this way, given its apparently successful record and its 1994 decision to continue its operation into the second half of the 1990s, but to renounce government money, which had previously made up about half of its funds.

Of course, even such a modest government role creates problems. First, it requires a precise definition in antitrust law, under which in precompetitive circumstances an R&D joint venture does not fall under the ban on collusion. In practice, any exemption clause will be abused to some extent, so the likely damage of abuse will have to be put onto the debit account of any (government-sponsored) cooperation. Second, it requires a decision as to what extent foreign firms are permitted to participate. Again, there is a conflict: on one side, the very rationale of technology policy is to foster national high-technology agglomerations, and not the ones in other countries; on the other hand, the participation of foreign firms may be the only feasible way to tap a foreign stock of knowledge that may be crucial to obtain or preserve the competitive edge of domestic industry. Even if foreign firms are excluded from participating in government-sponsored cooperation, however, they may acquire the relevant knowledge by taking over a firm that does participate.[34] In practice, this can hardly be avoided

---

[34] The history of SEMATECH is telling in this respect. See *The Economist,* 2 April 1994, "Uncle Sam's Helping Hand" (p. 93).

unless a government is prepared to make very serious inroads into the freedom of capital movements.

(3)   If an element of subsidization is to be added to the package of government R&D support, it should be done in a nondiscriminatory fashion. This means—roughly speaking—that any dollars or DM spent on R&D should be subsidized at the same rate, no matter which branch of economic activity it is invested in. This of course means that high-technology branches with high ratios of R&D spending to value added receive a higher subsidy per unit of value added; but this "discrimination" is perfectly compatible with the externality-based logic of the subsidy, because they are also the branches whose output is likely to be the furthest below the social optimum. In practice, a favorable tax treatment of R&D may be the most appropriate tool to achieve this task.

Of course, one may wonder whether most industrialized countries—including the United States and Germany—do not already have an implicit R&D subsidy implemented in their tax codes, although maybe for reasons that have nothing to do with a conscious and deliberate effort to support R&D. An implicit R&D subsidy can be recognized in two distinct elements of the tax code.[35] First, labor costs incurred in R&D can be deducted as current expenditure, like labor costs incurred in production, although they are economically more like an investment in future knowledge creation and thus, in a neutral tax code, would have to be treated like an investment in physical capital, which can be deducted only over time, according to some schedule of depreciation. Second, the value of the knowledge output that is not sold in the market for patents and licenses, but kept for exclusive use in the company, is not counted as an asset for tax purposes. This amounts to a positive discrimination of the knowledge stock vis-à-vis that part of the physical capital stock that is produced by the firm itself. While these tax rules may not be optimal from an economic perspective, they do constitute a considerable element of R&D support and maybe a starting-point for a "non-mercantilistic technology policy."

---

[35] For details, see M. Stolpe, "Industriepolitik aus Sicht der neuen Wachstumstheorie," *Die Weltwirtschaft*, 3 (1993) pp. 369–370.

Approaches to Conflict Resolution: Agenda for Action

SUBSIDIES AND PUBLIC PROCUREMENT

Toward an Open Subsidy Club[36]

Subsidies are one of the most important instruments for the promotion of high-technology industries. The share of public funding in private R&D expenditures varies between 11 and 33 percent in leading OECD countries, with the remarkable exception of Japan, where public subsidies cover only 1.5 percent of private R&D outlays (Bletschacher & Klodt, 1992). What is more, R&D subsidies are directed toward very few high-technology industries—above all to the electronics and aerospace industries—and often go along with investment and production subsidies, as the Airbus case demonstrates.[37]

In the course of the Uruguay Round negotiations, the GATT signatories have made a new attempt to restrict the use of subsidies for industrial policy purposes. The Agreement on Subsidies and Countervailing Duties introduced a so-called traffic light approach that divides subsidies into three categories: (1) prohibited; (2) actionable (i.e., these subsidies can be countervailed); and (3) nonactionable. The "red light" group includes export subsidies and subsidies that can be roughly characterized as import substitution subsidies.[38] These subsidies are actionable in any case, regardless of whether they are specific or not. The "green light" (nonactionable) category covers nonspecific subsidies, R&D subsidies, regional subsidies, and environmental subsidies. A subsidy is considered specific if it is granted to only a limited number of well-defined industries or enterprises. Although most R&D subsidies are designed to promote specific industries, R&D subsidies are non-actionable as long as public funds cover not more than 75 percent of the costs of industrial research and not more than 50 percent of precompetitive development research. To benefit from non-actionability, "green light" subsidies

---

[36] This chapter is based on J. Stehn, *Subsidies, Countervailing Duties, and the WTO: Towards an Open Subsidy Club,* Kiel Discussion Paper, Kiel Institute of World Economics, Kiel, June 1996.

[37] It has been calculated that almost one-third of all subsidies to Airbus Industries can be characterized as production subsidies. See G. Bletschacher, and H. Klodt, *Strategische Handels- und Industriepolitik,* Kiel Studies 244, Tübingen, 1992, pp. 64–66.

[38] "... subsidies contingent, whether solely or as one of several other conditions, upon the use of domestic over imported goods." (WTO Subsidies Agreement, Article 3.1.)

are to be notified to and approved by the WTO Committee on Subsidies and Countervailing Measures. All subsidies that fall into neither the "red light" nor into the "green light" category are defined as actionable ("yellow light" category).[39]

Although the WTO Subsidies Agreement clarified some of the issues that caused frequent disputes (Schott, 1995),[40] it suffers from at least two major weaknesses that might give rise to further conflicts in international trade, especially with a view to high-technology competition. First, countervailing duties imposed by third countries are the only enforcement mechanism of both the WTO Subsidies Agreement and the Tokyo Round Subsidy Code. To initiate a countervailing duty investigation a signatory has to prove that subsidies of an offended trading partner violate existing WTO regulations and that the domestic industry is "experiencing injury" as a result of these subsidies. Thus, industries and enterprises suffering from subsidies abroad are the real supervisors of the WTO subsidy regulations. This "decentralized" supervision system has proved to be rather ineffective on several grounds:[41]

• Many kinds of subsidies, especially investment subsidies and R&D subsidies, do not cause measurable competition distortion effects for several years. By the time the effects become obvious, the relevant subsidies have often been phased out, and it may become almost impossible to prove that the subsidies caused material injury.

• The current subsidy regulations cannot prevent governments from granting subsidies that distort international competition because a proof of material injury is a first and indispensable step in a countervailing duty investigation. Hence, countervailing duties can mitigate the competition distortions only after they have occurred.

• The countervailing duties approach poses the danger of retaliation, especially in high-technology industries, where the time lag between subsidies and the resulting competition distortion effects is usually quite long.

---

[39] According to the WTO Subsidies Agreement, both "yellow-light" and "red-light" subsidies are actionable. The main difference between these two groups is that "red-light" subsidies can—as a general rule—be countervailed by a foreign government without a proof of material injury.

[40] J.J. Schott, *Dispute Settlement in the Multilateral Trading System and High Technology Trade.* Paper presented at the conference "Towards a New Global Framework for High-Technology Competition," Kiel Institute of World Economics, Kiel, Germany, 30–31 August 1995.

[41] See L.D. Tyson, *Who's Bashing Whom? Trade Conflict in High-Technology Industries,* Washington, D.C., 1992, pp. 280–286, for a detailed evaluation of the Tokyo Round subsidy regulations.

Under these conditions, the imposition of a countervailing duty by a trading partner might be judged arbitrary and unfair by the offended country.

• Countervailing measures may be abused for protectionist purposes and—because of their price effects—may further distort international competition.

In view of these shortcomings, the WTO Subsidies Agreement should be reformed (1) by introducing a notification system aiming at an assessment of all planned subsidies prior to their implementation and (2) by defining quantitative thresholds for the provision of subsidies.

(1)   With respect to the implementation of a multilateral notification system, the aid supervision procedure of the European Union can serve as a reference system.[42] A multilateral subsidy supervision should stipulate that all plans to grant new or to alter existing subsidies are to be notified to and approved by the WTO Committee on Subsidies and Countervailing Measures (CSCM).[43] The CSCM should be entitled to examine the notified plans and to decide whether they are compatible with the WTO Subsidies Agreement. In the course of the investigation, the CSCM should take into account written comments by third signatories that might be affected by the notified subsidy. After the CSCM has made its decision, any signatory concerned should have the opportunity to initiate a panel procedure against the CSCM ruling in accordance with the WTO Dispute Settlement Mechanism. Given that a signatory grants a subsidy in violation of a final CSCM or WTO panel ruling, the CSCM should be empowered to require a repayment of the subsidy to the respective national government. The potential threat of a repayment may give an incentive to recipient firms and industries to ask their respective national governments to present an approval of the CSCM before granting the subsidy, and may, therefore, lead to some sort of self-restraint. Only if a signatory does not react to any of the CSCM or WTO panel rulings within an appropriate period (e.g., two months after the final decision), third parties would be entitled—as a last resort—to initiate a countervailing duty procedure according to the regulations of the current WTO Subsidies Agreement.

To facilitate a multilateral aid supervision, the current traffic-light approach should be reformed by categorizing all subsidies as either prohibited subsidies or subsidies that are allowed under certain conditions. This can easily be done by just skipping the "yellow light" category. Although this category is defined by default, it is obvious from the definitions of the "red light" and "green light" subsidies that it covers all specific subsidies not explicitly mentioned in the "green light" group. As a consequence, all

---

[42] See J. Stehn, "Wettbewerbsverfälschungen im Binnenmarkt: Ungelöste Probleme nach Maastricht," *Die Weltwirtschaft* (1993), pp. 43–60, for an evaluation of the European aid supervision system.

[43] Currently, only "green light" subsidies have to be notified to the CSCM.

non-specific subsidies except for the current "red light" group, i.e., export subsidies and import substitution subsidies, would be admissible. Thus, a workable and clear-cut definition of specificity is an important prerequisite for an effective subsidy supervision.

(2) The WTO Subsidies Agreement fails to provide unequivocal guidelines for measuring specificity. According to the agreement (Art. 2.1(c)), the following factors should be considered in determining whether a subsidy program is specific or not:

- use by a certain number of enterprises;
- predominant use by certain enterprises;
- the grant of disproportionately large amounts of subsidy to certain enterprises; and
- the manner in which discretion is exercised by administering authorities.

Besides being obviously rather vague and ambiguous and thus giving rise to disputes in interpretation, these guidelines do not take sufficient account of the economic effects resulting from firm-specific or industry-specific subsidies. From an economic point of view, the main objective of specificity rules is to limit the competitive distortions due to subsidies that are mainly directed to a single industry or enterprise. It can be realistically assumed that the extent of competition distortions depends on the share of production, investment, or R&D costs that is covered by public funding. Hence, the current rules could be considerably improved by defining quantitative thresholds that limit the provision of subsidies to a certain fraction—say, 5 percent—of the respective subsidy base.

With a view to R&D subsidies, it is certainly by far too optimistic to expect that governments will agree on a 5 percent threshold in the near future. The "safe harbour provision" (Schott, 1995)[44] for R&D subsidies which was agreed upon in the Uruguay Round negotiations rather points in the opposite direction. It is obvious that the current maximum, i.e., a public funding of up to 75 percent of R&D costs, will give a recipient firm a considerable competitive edge and thus might lead to major conflicts in high-technology trade and competition. Under these conditions, the levelling-down of current thresholds seems to be a necessary and highly desirable policy goal.

As empirical research indicates, high-technology R&D can be expected to generate cross-border externalities.[45] Thus, new technological knowledge can no longer be assumed to be totally exclusive. If national govern-

---

[44] J.J. Schott, *Dispute Settlement in the Multilateral Trading System and High Technology Trade.* Paper presented at the conference "Towards a New Global Framework for High-Technology Competition," Kiel Institute of World Economics, Kiel, Germany, 30–31 August 1995.

[45] See above (Analytical Benchmarks).

ments were to take these international spillovers of domestic innovations into account, they would have strong incentives to cooperate internationally.[46] An efficient way of international cooperation might be the mutual opening up of national subsidy funds for high-technology development. This approach would allow firms located in third countries open access to national subsidy funds if—and only if—third countries are ready to offer open access to their subsidy budgets on a reciprocal basis.

To provide an incentive to open up national subsidy funds, one could also consider a provision that, as a general rule, lowers the threshold for high-technology subsidies to 30 and 15 percent of the costs of basic and applied R&D, respectively, but allows higher public funding—up to the current limits—if a national research program provides open access for firms located in third countries. For practical purpose, this approach may require the implementation of the following general rules:

- An R&D subsidy programme should be regarded as open only if at least two foreign firms are participating in the program;
- the objectives of the open fund are formulated by the respective national government and must be met by both domestic and foreign firms;
- domestic and foreign firms should be treated equally with respect to patents and copyrights that emerge from the funded research; and
- open access to national funds should not be linked to cooperation with a domestic firm.

To facilitate a multilateral subsidy supervision and international cooperation in publicly funded R&D, the WTO should publish annual reports on recent developments in national subsidy schemes of leading OECD countries. Multilateral monitoring of current subsidization practices could also help to prevent international conflicts arising from a misinterpretation of the objectives and potential effects of national subsidy programs (Ostry, 1995).[47]

Public Procurement:
Conflict Resolution by National Courts

Public procurement in advanced industrial economies covers a significant part of overall market demand. Non-defense public procurement in the member states of the European Union is estimated to represent about 7 to

---

[46] In an extreme case, this could lead to a free-rider problem in promoting private R&D. However, with a view to the fact that the number of players in the high-technology game is rather small, this seems to be very unlikely.

[47] S. Ostry, *Technology Issues in the International Trading System,* September, 1995, unpublished paper.

10 percent of gross domestic product.[48] It is therefore tempting for governments to use their demand for goods and services to achieve aims of technology and industrial policy.

The GATT procurement code of the Tokyo Round (in force since 1981) established for the first time internationally binding rules that secured foreign competitors open and undiscriminatory access to bidding procedures for public procurement of goods (though not yet of services).[49] It is based on the conditioned most-favored-nation clause, i.e., it applies only to the relationships between those contracting parties of GATT that have actually signed the code.

The Tokyo Code, at that time widely regarded as one of the most far-reaching achievements of the Tokyo Round Agreements,[50] committed only central governments and directly related entities to internationally open tenders (i.e., those surpassing certain threshold levels of tender value). The rather limited range of the Tokyo Code—covering less than 10 percent of nondefense public procurement in the United States and the European Community[51]—clearly constrained its impact from the outset and goes a long way toward explaining its actually very limited economic consequences. As a part of the Uruguay Round accord, the new Government Procurement Agreement (GPA)—in force from 1 January 1996—extends the reach of the Tokyo code to potentially all kinds of nondefense procurement, i.e., to all nondefense goods and services.

The GPA is one of the three major agreements of the Uruguay Round that are not included in the so-called Single-Undertaking procedure. This means that these agreements apply only to their signatories, not to all contracting parties of the World Trade Organization at large.[52] The potential

---

[48] See Patrick A. Messerlin, "Agreement on Public Procurement," in OECD, *The New World Trading System:* Reading, OECD Documents, OECD, Paris, 1994, pp. 65–71. Figures of this kind of course depend on the underlying definition of what is considered public sector procurement. This concerns in particular the procurement policy of government-owned firms and the procurement policy of firms operating in markets that are heavily regulated by government (such as utilities).

[49] Government procurement had been previously exempted from GATT regulations (Art. III (8) GATT).

[50] See Robert Stern, and Bernard M. Hoekman, "The Codes Approach," in J. Michael Finger and Andrzej Olechowski (eds.), *The Uruguay Round: A Handbook on the Multilateral Trade Negotiations,* The World Bank, Washington, D.C., 1987.

[51] See Patrick A. Messerlin, "Agreement on Public Procurement," OECD, *The New World Trading System: Reading,* OECD Documents, OECD, Paris, 1994, pp. 65–71.

[52] The other two agreements are the Arrangement Regarding Bovine Meat and the International Dairy Arrangement. A fourth plurilateral agreement under the umbrella of the WTO, the agreement on Trade in Civil Aircraft, was not changed at the end of the Uruguay Round and remained open to signature only in its already existing form. Parties of the GPA are at present Canada, the fifteen member states of the European Union, Norway, Switzerland, Japan, the United States, Israel, and South Korea. Singapore was a signatory of the Tokyo Code, but opted out this time. In turn, South Korea entered the club as a new member.

value of contracts that will be covered by the GPA in signatory states is roughly estimated to be about US $ 400 billion annually in current prices.[53]

There are two main areas in which the GPA breaks new ground by the standard of its Tokyo Code predecessor. These concern (1) extensions of the coverage provided by the GPA and (2) disciplines imposed by the GPA on signatories.

(1) *Coverage of the GPA:* The principle of exchange of "trade concessions" that has governed all multilateral trade negotiations under the auspices of GATT means in the context of GPA that contracting parties concede "items" (i.e., certain groups of goods and services) to be opened to foreign competition in public tenders, and "entities" (public or semi-public bodies) designated to open their tenders for bids by foreign suppliers.

The new GPA extends the range with respect to both "entities" and "items" concerned. Under the new agreement (Article I), all tenders regarding rentals or leases of goods, as well as tenders regarding the procurement of services, shall be included.

These general extensions are, however, limited by exemptions listed in the Annex of the Agreement for each signatory state.[54] Limits concerning goods are listed as exemptions; limits concerning services apply to all items that are not explicitly enumerated. This distinction is in accordance with the different philosophies of regulation in the GATT (for goods) and the GATS (for services). The institutional solutions chosen in service sectors are likely to favor bilateral deals of sectoral reprocity, thus undermining attempts to arrive at a multilateral framework. A similar tendency prevails in the so-called exempted areas of public procurement of goods, notably telecommunications equipment. As far as public or semi-public "entities" are concerned, the new GPA has in principle been extended to "sub-central" entities, i.e., basically regional and local government entities. However, those sub-central entities that are in fact obliged to open their tenders to foreign competition are enumerated in the annex to the GPA.[55]

There are two major problems involved in the definition of "public entities." First, the power of the contracting party—the central government—to

---

[53] See Jeffrey Schott (assisted by Johanna W. Buurman), *The Uruguay Round: An Assessment.* Institute for International Economics, Washington, D.C., November. 1994. OECD, Trade and Competition Policies: Comparing Objectives and Methods, Trade Policy Issues, 4, OECD, Paris, 1994. Sylvia Ostry and Richard R. Nelson, *Techno-Nationalism and Techno-Globalism: Conflict and Cooperation,* Brookings Institution, Washington, D.C., 1994.

[54] Note that the public procurement regulation of the European Union, which in many instances seems to have served as an example for the WTO procurement regulation, does not contain exemptions by "public entities" or "items" except the general exemption of defense procurement.

[55] For instance, in the case of the United States, many of the obligations under this treaty are limited to a rather small number of states.

control the conduct of sub-central entities may be rather limited, e.g., in the case of federal member states. Second, previously "public firms," once included in international arrangements on public procurement, may become privatized, thus leaving other contracting parties with a loss of "trade concessions" formerly granted to them.

(2) *Disciplines Imposed by the GPA:*[56] According to the two basic principles embedded in the GPA, suppliers from other signatory states should benefit from the conditional MFN clause, and thus tendering procedures should not entail any discrimination between domestic suppliers and suppliers from other signatory states. In turn, discrimination against suppliers from third countries (i.e., nonmembers of the GPA) is still allowed, which is consistent with the philosophy of the WTO as an open club.

A core achievement of the new GPA is that foreign suppliers discriminated against in a national tendering process can use the so-called challenge procedure, i.e., they can submit their appeal directly to a ruling by the courts of the country that issued the respective tender, and these courts are then obliged to provide reasonably rapid proceedings.[57]

Moreover, "offsets" (i.e., deviations from the GPA) are explicitly prohibited by Art. XVI, meaning that additional requirements attached to a bid on a national public tender (e.g., local content, countertrade, and the like) are to be considered illegal. Furthermore, the GPA does not contain any safeguard clause that could allow signatory states to refrain from fulfilling, or to circumvent, their obligations under this agreement. On the other hand, the GPA does not contain any provision against collusion among bidding domestic firms, which is left to the competence of domestic competition policy.

The regulations of the GPA, which by themselves appear to be quite strict,[58] may yet be undermined by regulations of the TRIPs Agreement

---

[56] Cf. Bernard M. Hoekman and Petros C. Mavroidis, "The WTO's Agreement on Government Procurement: Expanding Disciplines, Declining Membership?" Discussion Paper 1112, CEPR, London, 1995.

[57] In addition, the WTO dispute settlement procedures are open to such cases and can also be used. There are, however, significant differences between these two routes of appeal against discriminatory treatment by particular national "public entities." To be successful in a WTO dispute settlement procedure, the claimant (the government of the affected firm) must provide evidence that its previously granted trade concessions were impeded in the case in question.

[58] Here are selected articles of the GPA with some relevance to the issue of maintaining club discipline: Art. VI defines rules for technical specifications of the items included in a tender; Art. VII regulates the choice among possible tendering procedures; Art. VIII is concerned with qualification requirements for tenderers; the invitation procedures to participate in a tender are circumscribed in Art. IX; the selection procedure is outlined in Art. X; prescriptions on time schedules are given in Art. XI and on necessary documentations in Art. XII; submission procedures are regulated in Art. XIII; Arts. XIV through XVIII are primarily

(Agreement on Trade-Related Intellectual Property) of the Uruguay Round. This means in particular that a national government will still be allowed to specify the conditions of a tender, if "there is no sufficiently precise or intelligible way of describing the procurement requirements and provided that words such as "or equivalent" are included in the tender documentation" (Article VI.3 GPA).[59] This provision may give national governments enough leeway to limit international competition whenever they think it to be appropriate.

In practice, foreign firms affected by discrimination in public tendering procedures will weigh the possible advantages of a court ruling in their favor against the disadvantages of possibly foregoing the good will of the government concerned. In the longer term, however, the opportunities established for foreign suppliers' access to domestic court procedures should work in the direction that national governments will increasingly behave more strictly according to the rules laid down in the GPA.

Although clearly the GPA is a major step in the right direction, major problems remain unsolved:

• At present, WTO regulations still allow legal subsidies to R&D activities and subsidies on regional policy grounds. This opens the gate for abuse in national procurement: e.g., a national government may grant R&D subsidies and then define the terms of a later procurement tender on the basis of those specific R&D requirements that only a domestic firm can possibly meet. Bidders from the outside would obviously have little chance to succeed in such a procurement tender.

• Decisions of governments on matters of national security are exempted from WTO regulations. Obviously, clauses of this kind can also be misused, in particular with respect to the treatment of dual-use goods and services (i.e., goods and services that can be used for both military (security) and civilian purposes).

• The many national exemptions in the GPA with respect to public entities and items (or sectors) are likely to encourage bilateral bargaining, which is obviously not in the spirit of an international and multilateral framework of trade regulations.

• The question of what has to be considered a public enterprise or an enterprise under significant influence of government, can hardly be an-

---

concerned with other technicalities of the tendering process; and finally, Art. XIX requires that the parties concerned collect annual statistics and provide these statistics to the Committee on Government Procurement at the WTO—which by the way can be considered the "nucleus" of a future supervision board in this area, the creation of which is considered necessary by many experts (see Sylvia Ostry, "Technology Issues in the International Trading System," unpublished paper, OECD Trade Committee, September, 1995.)

[59] See also Jean-Jacques Laffont and Jean Tirole, "Auction Design and Favoritism," *International Journal of Industrial Organization,* vol. 9, 1991, no. 1, pp. 9–42.

swered on the basis of common criteria across countries. As this definition is critical for the assessment of trade concessions in international trade negotiations, there is a strong need for further international agreements on this matter.

As the new GPA code is entering into force by 1 January 1996, it remains to be seen how effective this code will prove to be, and whether the problems enumerated above will in fact call into question the overall spirit of the whole venture.

MARKET ACCESS AND STRUCTURAL IMPEDIMENTS

In addition to subsidies and public procurement practices, international competition in high-technology goods is distorted by a variety of other barriers to market access, including "structural" impediments to trade and investment. Trade and investment are in fact two major channels for the international diffusion of technology. In exporting countries, the foreign trade outlet enables technology enterprises to exploit economies of scale in production and facilitates their recovery of R&D expenditures. In importing countries, user industries benefit from access to the most advanced technology-intensive capital goods, components, and services at competitive prices, often a precondition for their own international competitiveness. Consumer welfare, too, is enhanced by the supply of high-technology consumer goods and services, again at competitive prices. From this one might infer that governments should be interested in open markets for high-technology goods and services.

In practice, however, barriers to market access prevail. Some barriers, like tariffs, certain antidumping practices, voluntary export restraint agreements, and safeguards, are deliberate attempts to shield domestic producers against competition from technologically superior, or simply less expensive, foreign suppliers, often with the intention of providing them a respite for catching up. Others, like government regulatory measures and technical standards, are often the by-product of differences in national cultures, policy objectives, and technology practices which may be difficult to reconcile internationally. And still others, like barriers to access to private technology "clubs" or to national dealer networks, are the result of restrictive, though not necessarily illicit, business practices. All these practices are potential sources of economic and political friction.

Tariffs and Non-Tariff Barriers

Tariffs, traditionally the preferred instrument to restrict market access, have greatly lost importance in the course of repeated GATT rounds. The

Uruguay Round resulted in a (further) lowering of import duties for industrial goods by 38 percent in developed countries, from 6.3 to 3.9 percent on average. Forty-three percent (up from 20 percent) of the total value of industrial products imported by developed countries are now entering duty-free, and only 5 percent (down from 7 percent) are subject to peak rates.[60] Equally important as the reduction of tariffs is that developed economies have now accepted a binding of virtually all their tariff lines on industrial products, thereby greatly improving the security of market access.[61] Technology products, among them electrical and nonelectrical machinery, chemicals, and pharmaceuticals, were also subject to the elimination or significant reduction of tariffs; on important electronics items, such as semiconductors, semiconductor manufacturing equipment and computer parts, tariff cuts of 50 percent and above were achieved. Yet, for certain items, especially in the field of consumer electronics, substantial tariff barriers persist. Their reduction should be made part of a concerted multilateral effort to improve market access in which the European Union, the United States, and the major East Asian economies should take the lead.

With MFN tariffs rendered increasingly obsolete as measures of protection, at least in developed economies, non-tariff barriers to trade and market access have received growing attention. Among *border measures*, anti-dumping duties, countervailing duties, and safeguards are increasingly applied in high-technology trade. They are treated in the following chapter. The much criticized Voluntary Export Restraints (VER), a frequently applied substitute for these measures, shall be phased out as a result of the Uruguay Round. No less significant than border measures are the already mentioned structural impediments *inside the borders:* government regulatory practices, technical standards, and restrictive business practices. They may be deliberately designed in such a way as to discriminate against imported products: luxury taxes on certain classes of automobiles, the U.S. corporate average fuel economy (CAFE) law,[62] costly and time-consuming testing, certification, and conformity assessment procedures for technology-intensive products, and limitations to foreign ownership in business sectors deemed "strategic." Discriminatory measures are an offense against Article III of the GATT, which establishes the principle of national treatment (non-discrimination) of imported products, a guiding principle of the multilateral trading system. Such measures are therefore challengeable under the GATT

---

[60] Cf. GATT, News of the Uruguay Round, April 1994.

[61] The percentage of tariff lines bound has been increased from 78 to 99 percent in developed economies, and from 22 to 72 percent in developing economies. Ibid.

[62] Cf. G. Kleinfeld, "Taxation of Automobiles and GATT National Treatment Obligations: Where to Draw the Line?" *World Competition,* vol. 19 (1995) no. 1, pp. 77–90.

dispute settlement mechanism. Often, however, structural impediments simply reflect differences in national value systems, historic institutional settings, and grown systems of business culture (like the systems of corporate governance). While they may form effective barriers to foreign entry, the WTO recognizes the diversity of national business cultures, institutional settings, and policy approaches which are at the heart of comparative advantage[63] and competition among which is indeed a condition for advancement. It is no wonder, then, that the Uruguay Round has made only limited progress in this difficult area. To be sure, the Agreement on Technical Barriers to Trade, concluded under the Tokyo Round, was revised and strengthened, and Article XXIII of GATT was clarified. What is still missing is, among others, an agreed catalogue of government practices (including omissions) considered to be (challengeable) structural impediments, as well as "objective" criteria for determining whether and to what extent a market is indeed "contestable,"[64] and if not, how this situation can be remedied.

In the field of high technology, market presence through foreign direct investment is no less important than market access through trade. Domestic presence is often critical to success in following market trends, gaining access to the R&D infrastructure, marketing research-intensive products with short product cycles and high sunk costs, and responding to marketing needs. At the same time, foreign investment is an important channel for the international diffusion of technology, including skills. In spite of these microeconomic and macroeconomic benefits, the notion of national treatment, while accepted in principle, is not yet fully applied to investment.

Governments seek to attract greenfield investment from foreign high-technology enterprises through a great variety of financial incentives and/or through trade pressure. At the same time that they restrict market presence in business sectors deemed "strategic," they oppose the acquisition of domestic technology firms by foreign investors, or they restrain access to national technology programs. True, an opening of the respective sectors to foreign investment often goes hand in hand. E.g., in telecommunications, a high-technology service sector, deregulation in Europe has promoted foreign entry as well as the formation of international strategic alliances with non-European partners. Still, in all countries of the triad, and even more in emerging and developing economies, obstacles and market distortions persist, as foreign investors are denied effective market presence.

The Uruguay Round Agreement on Trade-Related Investment Measures (TRIMs) is but a first step to a multilateral investment agreement. The

---

[63] Cf. S. Ostry, "Technology Issues in the International Trading System," in OECD, *Market Access after the Uruguay Round: Investment, Competition and Technology Perspectives,* Paris, 1996, p. 159.

[64] Ibid.

agreement is narrow in its focus, being limited to the trade-related aspects of investment, and here in particular to local content and trade balancing requirements, which are expressly prohibited. This prohibition applies whether the measures are mandatory or are required in return for an incentive or advantage. The TRIMs Agreement had a poor start: under the U.S.-Japan Auto Agreement of 1995, the Japanese car manufacturers entered into a "voluntary" commitment to raise the U.S. content of their transplants in the United States and to meet NAFTA local content standards by 1998 "in order to boost U.S. and North American content in their vehicles."[65] This commitment, like the other commitments under the agreement, will be monitored by "a new, effective monitoring system jointly developed by the U.S. industry working closely with the U.S. government."[66] It remains to be seen whether the auto agreement will serve as a model for circumventing the TRIMs Agreement by "voluntary" commitments or whether it remains a one-time transgression.

What is needed beyond the TRIMs Agreement is a General Agreement on Investment (GAI), the scope of which is not limited to trade-distorting measures. This agreement should guarantee the right of establishment and full national treatment,[67] subject only to exceptions which are clearly and narrowly defined (e.g., national security) and open to judicial review.[68] Among the issues to be addressed under a GAI should be: access to government technology-support programs; state sanctioned monopolies and other sector-specific reservations; most-favored nation (MFN) treatment; transparency; barriers to foreign takeovers; performance requirements; investment incentives; restrictive business practices and competition policy; investment protection; access to technology; payments and transfers; movement of key personnel and data; and a dispute settlement mechanism for resolving conflicts not only between governments, but between governments and investors.[69]

---

[65] USIS, "U.S. government fact sheet: U.S.-Japan Auto and Auto Parts Agreements," *U.S. Information and Texts,* 30 June 1995, p. 11.

[66] Ibid.

[67] Cf. R.Z. Lawrence, "Towards Globally Contestable Markets," in OECD, *Market Access after the Uruguay Round: Investment, Competition and Technology Perspectives,* Paris, 1996, p. 29.

[68] Cf. E.M. Graham, "Investment and the New Multilateral Trade Context," in OECD, *Market Access after the Uruguay Round: Investment, Competition and Technology Perspectives,* Paris, 1996, pp. 4f. The authority granted to the U.S. president under the Exon-Florio amendment to block the takeover of a U.S. enterprise by a foreign investor in the case of an "impairment of national security" is much too general, and it is not subject to court challenges.

[69] Cf. Ibid., pp. 35–62; D.M. Price, "Investment Rules and High Technology: Towards a Multilateral Agreement on Investment," in OECD, *Market Access after the Uruguay Round: Investment, Competition and Technology Perspectives,* Paris, 1996, pp. 171–186; D.L. Aaron, "After the GATT, U.S. Pushes Direct Investment," *The Wall Street Journal,* 2 February 1995.

Work for a Multilateral Agreement on Investment (MAI) is currently under way in the OECD. This takes account of the fact that most of the issues at stake are less controversial among industrialized countries than with emerging and developing economies. Nevertheless, the aim must be for an agreement under the WTO to complement the GATT and GATS Agreements and covering these countries as well, though initially a WTO agreement may be less far-reaching in scope and commitments than a MAI under the OECD.

In the face of what has been perceived as "inadequate" access to markets due to structural impediments, *effective market access* has become the catchword for triggering (and/or rationalizing) unilateral trade measures aimed at forcing open foreign markets and securing targeted market shares for imported (technology) products, thus circumventing the laborious procedure of negotiating away specific obstacles. This was especially the U.S. approach toward Japan under (the threat of) Section 301 in the cases of semiconductors, flat glass, telecom equipment, medical equipment and, recently, autos and auto parts. The European Community has as yet made little use of its New Commercial Policy Instrument (NCPI) introduced in 1984 or its successor, the Trade Barriers Regulation (TBR). Between 1984 and 1994 only four examination procedures were opened. In one case the EU obtained a favorable GATT decision; in the three other cases the disputed practices were discontinued without the EU taking retaliatory measures.[70] With access of firms to the new TBR, "liberalized" relative to the old NCPI, "it is likely that the use of the instrument will be more frequent in the future."[71] In the face of effectively closed markets, with no clearly identifiable—and yet quite effective—barriers to entry, taking recourse to market-opening instruments such as Section 301 or the Trade Barriers Regulation appears at first sight an appropriate way of approaching the obstacles. The temptation of a unilateral approach is all the greater since experience in the semiconductor and automobile cases, and in cases which were resolved with less publicity under the threat of these instruments, may well lead to the conclusion that this is indeed a successful means of enforcing market access.

Yet there are serious arguments against a "managed trade" approach, especially if applied unilaterally. First, unilateral action undermines the multilateral trading system which is an international public good. Secondly, a unilateral approach lends itself only to countries or trading blocs with a high bargaining leverage, such as the United States and the European Union, with the rest of the world put at a clear disadvantage. It is therefore

---

[70] Cf. H. Beekmann, "The 1994 Revised Commercial Policy Instrument of the European Union," *World Competition,* vol. 19 (1995) no 1, pp. 59f.

[71] Ibid., p. 75.

liable to misuse as an instrument of power politics. Thirdly, because of that there is a strong presumption that any concession achieved will be at the expense of third parties. This is even true if the claimant pretends not to seek preferred treatment over other foreign supplier countries, since the defendant country may not be prepared to open its market for imports generally and may well curtail existing market shares of third parties in favour of the economic superpower involved. Moreover, follow-up measures are discussed bilaterally without the participation of third parties.[72] Fourthly, lack of market access is often due to one's own insufficient marketing effort. Third-country suppliers which undertook such effort at high cost will find their investment depreciated by the easy access offered to their foreign competitors as a result of government pressure.[73] Fifthly, more often than not quantitative import targets can only be met through active involvement and commitment of domestic suppliers in the country concerned. This encourages collusive agreements among these firms and furthers the cartelization of the respective market, a result which is counterproductive to economic efficiency.

With the dispute settlement procedure greatly improved as a result of the Uruguay Round, there is therefore a strong case for taking resort to multilateral rather than unilateral action. Article XXIII:1(b), while not yet tested in practice, may offer access to the GATT dispute settlement procedure in cases where a contracting party finds that "any benefit accruing to it directly or indirectly under this Agreement is being nullified or impaired or that the attainment of any objective of the Agreement is being impeded as the result of the application by another contracting party of *any measure, whether or not it conflicts with the provisions of this Agreement*"[74] ("non-violation clause"). The Articles of Agreement of GATT thus seem to offer a multilateral route to dispute settlement even in cases of "structural" impediments, including private restraints of trade.[75] This route should be tested and, if found inadequate, be improved.[76]

---

[72] G. Zeidler, "Distribution Systems and Vertical Integration: Protectionism or 'Normal' Business Practice?" Discussion paper prepared for the conference "Towards a New Global Framework for High-Technology Competition," held in Kiel, Germany, on 30–31 August 1995.

[73] For instance, U.S. car manufacturers "estimate" that under the U.S.-Japan Auto Agreement of 1995, Japan will increase its imports of Big Three U.S.-made vehicles from 45,000 in 1994 to 160,000 by 1998. This "estimate" is the basis for the monitoring of the agreement. A similar "goal" is the establishment of 200 new dealerships in Japan by 1996 and 1,000 by the year 2000 which is based on access to dealership networks of Japanese auto manufacturers. Cf. "U.S. Government Fact Sheet: U.S.-Japan Auto and Auto Parts Agreement," *U.S. Information and Texts,* 30 June 1995, p. 11.

[74] Italics added.

[75] Cf. S. Ostry, "Technology Issues in the International Trading System," in OECD, *Market Access after the Uruguay Round: Investment, Competition and Technology Perspectives,* Paris, 1996, pp. 160f.

[76] As Ostry remarks, "The 1995 auto dispute between the United States and Japan was a lost opportunity to launch that process." Ibid., p. 161.

Technical Norms and Standards

Major issues concerning market access are technical norms and standards. Generally speaking, standards, whether aimed at social objectives like human health and safety (regulatory standards) or at reducing transaction and information costs (compatibility standards), can be considered as welfare enhancing.[77] Yet, they may also serve as instruments to promote the international competitiveness of national industries, and they may act as barriers to trade and market access. In both cases they give rise to friction. In consideration thereof, in the Tokyo Round the Agreement on Technical Barriers to Trade (TBT) was concluded. In the Uruguay Round this Agreement was revised and strengthened in several respects. The Agreement now applies to all member countries of the WTO rather than to a limited number of signatories only, and it now covers not only the central government bodies but, though with major reservations, local government and non-governmental bodies, too. In the latter cases the member countries "shall take such reasonable measures as may be available to them to ensure compliance" with the provisions of the Agreement.

The Agreement establishes national treatment and MFN status as the guiding principles. It obliges the member countries to ensure that technical regulations as well as conformity assessment procedures are not prepared, adopted or applied with a view to or with the effect of creating *unnecessary obstacles* to international trade. The key concept of unnecessary obstacles has been clarified to the effect that regulations and procedures shall not be more trade-restrictive than necessary to fulfil a *legitimate objective,* taking account of the risks non-fulfillment would create. Among such legitimate objectives are national security requirements, the prevention of deceptive practices, protection of human health or safety; animal or plant life or health; or the environment. Where relevant international standards exist they shall be used as a basis for national technical regulations, except if this would be ineffective or inappropriate because of "fundamental" climatic, geographical or technological factors. Whereever appropriate, member countries shall specify technical regulations in terms of performance rather than design requirements. The Agreement sets forth provisions for the ex ante and ex post information on technical regulations ("transparency"); equitable procedures for conformity assessment, testing, and certification including information requirements, the respectance of confidentiality of information about products; equitable fees; recognition of conformity assessment in other member countries; and consultation and dispute settlement according to the rules of the WTO Understanding.

---

[77] Cf. B.M. Hoekman and P.C. Mavroidis, *International Antitrust Policies for High-Tech Industries?* Paper presented at the conference "Towards a New Global Framework for High-Technology Competition," held in Kiel, Germany, 30–31 August 1995, pp. 10f.

Standards are set by the government, private standards organizations, and individual enterprises or alliances. There are no clear-cut boundary lines between the three: the—now dropped—European HDTV norm was agreed between the European Community and leading European manufacturers. As major, and sometimes sole, buyers of specific technology products, large state enterprises or monopolies, such as railroads, post and telecommunications, and other public utilities, have a decisive influence on the relevant technical standards. Most of the standards are set by the industrial standards organizations. In Europe, eighteen national organizations, such as DIN in Germany (with about 25,700 standards in 1986), Afnor in France (with 13,400), and BSI in the United Kingdom (with 9,400),[78] work together in the Comité Européen de Normalisation, (CEN), the Comité Européen de Normalisation Electrotechnique (CENELEC), and the European Telecommunications Standards Institute (ETSI), reaching even beyond the European Union. Standards decisions are taken by qualified majority under a system of weighted votes.[79] In the United States, the process is more complicated. Approximately 400 organizations, working independently of one another and only part of them coordinating their activities in the American National Standards Institute (ANSI), are involved in standards development.[80] This puts obvious limits to any attempt to arrive at a binding plurilateral standards-setting procedure following the European model. On the international level, only seventeen out of 89,000 standards used by the United States in 1989 were adopted from ISO standards, a shortcoming which has been criticized as violating the GATT Standards Code.[81]

In standards enforcement, there are as many as 44,000 jurisdictions in the United States, resulting in "overlapping responsibility and redundant standards and regulations. In some cases, the products are regulated directly through inspection or testing programs, or both. In other cases, an approval body may have to certify that products meet standards set by a particular state or municipal government. This becomes a technical barrier in cases where states and municipalities have regulations that apply different standards, or where certification requirements differ."[82] When testing

---

[78] U.S. Congress, Office of Technology Assessment, *Global Standards: Building Blocks for the Future,* Washington, D.C., 1992, p. 63, based on: F. Nicolas and J. Repussard, *Common Standards for Enterprises,* Luxembourg, 1988, p. 26.

[79] U.S. Congress, *op. cit.,* p. 70.

[80] Ibid., pp. 49ff.

[81] B. Dhar, "The Decline of Free Trade and U.S. Trade Policy Today," *Journal of World Trade,* vol. 26 (1992), no. 6, pp. 149ff., based on Commission of the European Communities, *Report on United States Trade Barriers and Unfair Trade Practices,* Brussels, 1991.

[82] Department of Foreign Affairs and International Trade, *Register of United States Barriers to Trade, 1994,* Ottawa, 1994, pp. 11f., quoted in S. Ostry and R.N. Nelson, *Techno-Nationalism and Techno-Globalism: Conflict and Cooperation,* Washington, D.C., Brookings Institution, 1995, p. 94, footnote 15.

organizations even within the United States need multiple state and local accreditation,[83] is it conceivable that the principle of home country control be applied in trans-Atlantic (and/or trans-Pacific) trade? In the European Community, this principle ("country-of-origin" principle—mutual recognition of standards and procedures) is a cornerstone of the so-called New Approach to Standardization adopted as part of the Single Market program. It is coupled, though, with minimal harmonization of performance requirements and even, where necessary to ensure compatibility and interconnection, of design requirements as well as of testing and certification procedures. Under the New Transatlantic Marketplace, a key component of the New Transatlantic Agenda endorsed at the U.S.-EU Summit in Madrid on 3 December 1995, the United States and the European Union undertake to work toward agreements recognizing each other's testing and certification of products in a number of sectors, among them computers, telecommunications, and business equipment.[84] This is a first step in the same direction.

In new areas of high-technology, the problem of an international harmonization of compatibility standards is increasingly being "solved" by standards-setting by leading firms or strategic international alliances which seek to promote their own standards internationally, often in competition with rival groups, by-passing the traditional standards organizations. Among the examples are DOS vs. Apple and OS/2, VHS vs. Betamax and Video 2000, Hi-8 vs. S-VHS, and GSM vs. CDMA and PDC (the respective standards for mobile phones in Europe, the United States, and Japan). While the problem of international compatibility is solved to the extent that a *firm* standard is able to succeed as a new *industry* standard, issues of market power and collusive behavior, including "fair" access to the "club" for competing firms, may arise. This poses problems for international competition policy.

DUMPING, ANTIDUMPING AND COMPETITION POLICIES

Private versus Public Conduct in Trade and Competition

Apart from public obstacles to, and distortions of, high-technology trade and competition, trade-related barriers to market entry raised by private companies become increasingly important. This also means a greater role for competition policies and their international coordination. Existing international trading rules cover private business conduct to the extent that dumping practices are involved. An issue of particular interest in this context is the

---

[83] Ibid.

[84] USIS, "U.S.-EU Economic Relationship: the NTA Marketplace," *U.S. Information and Texts,* 7 December 1995, p. 9; USIS, "Joint U.S.-EU Action Plan," *U.S. Information and Texts,* 7 December 1995, p. 16.

question of whether antidumping policies should—and could—in future be replaced by genuine competition policies. Competition policy will also have to deal with private business practices which so far are not subject to international disciplines but may equally disrupt international trade and competition. These include various forms of international technological cooperation.

The dumping practices of companies and the antidumping policies of governments both lead to conflicts between countries as well as inside the borders. The international conflicts concern mainly producer interests. Foreign dumping causes friction if it hurts domestic industries. However, the policies applied in response to dumping are no less contentious among trading partners if weak domestic industries are protected against efficient foreign competitors. In the national context, friction arises if antidumping policies work to the detriment of consumers and user industries.

Defined as selling abroad at prices below "normal value," dumping is assumed to be unfair per se, regardless of its overall welfare impact. For protection measures to ensue, it is normally sufficient that dumping is shown to cause "injury" (e.g., losses to sales, employment, or profit) or to threaten to do so, to a domestic industry or to retard its establishment. Antidumping therefore has become the instrument of choice for industries seeking to reduce competition from imports. In many cases, it is "just ordinary protection with a good public relations program."[85]

In contrast to its actual "capture" by special interests, the economic rationale of antidumping policy (and indeed its original design in the United States in 1916 as an international extension of competition law to deal with cross-border violations of antitrust standards) is based on the existence of protected monopolistic market structures in foreign countries, facilitating "predatory" pricing strategies which finally lead to a dominant position of exporting countries on the importer's market. For predatory dumping to take place, the foreign monopolist (or cartel) must not only eliminate existing competition in the importing country, but must also be able to prohibit entry by new (domestic and foreign) competitors. It must either establish a global dominance or convince the host government to pursue a policy that tolerates or supports entry restrictions.[86] However, even the logical foundation of the argument appears weak. It seems more reasonable to argue that both the predator and the would-be victim would be better served by colluding with one another or by the takeover of the latter by the former than by going through the evidently expensive procedure of dumping.[87]

---

[85] J.M. Finger, *Legalized Backsliding: Safeguard Provisions in the GATT* [mimeo], Washington, D.C., 26 January 1995, p. 29.

[86] See B.M. Hoekman and P.C. Mavroidis, *Antitrust-Based Remedies and Dumping in International Trade*, August 1994.

[87] See P.K.M. Tharakan, "Anti-Dumping Policy and Practice of the European Union: An Overview," *Economisch en Sociaal Tijdschrift*, 1994, no. 4.

In actual fact, cases of successful predatory dumping remain largely undocumented. A recent study for the OECD concludes that "in the overwhelming majority of antidumping cases, there was little or no threat to competition in the domestic market."[88] This was true in more than 90 percent of the cases in which the United States and the EU imposed antidumping duties in the 1980s, and especially so with high-technology products such as semiconductors, compact disc players, and plain-paper copiers. In particular, no clear evidence was found that Japanese producers of electronic products applied "monopolizing" dumping designed to cripple foreign competitors.

Therefore, international dumping, as defined in antidumping regulations, largely appears to be normal competitive business behavior aimed at expanding foreign market shares, offsetting a sudden fall in demand, or exploiting economies of scale. It is international price differentiation or "penetration" pricing below average cost (and even below short-run "static" marginal cost, which can be significantly higher than long-run "dynamic" marginal cost) that is consistent with free trade, as it boosts competition to the benefit of the domestic economy as a whole. In consequence, the policies to curb "dumping" frequently harm the whole economy, as they pit the interests of a few, if powerful, domestic producers against those of the domestic "many" and of foreign competitors. A recent study by the U.S. International Trade Commission counts the net welfare loss for the United States of the 279 antidumping (and countervailing) duties that were in place in 1991 at roughly 1.6 billion dollars.[89]

Apart from the standard static welfare losses, antidumping measures have important dynamic consequences. Other than patent policies, which lower static welfare, but do so by increasing the long-run rate of growth, antidumping action, by isolating national markets, tends to slow the pace of innovation (and with it, growth). It may induce industrial users of the protected product to relocate plants outside the domestic territory, in order to avoid the duty, and in effect widen rather than close existing efficiency gaps between foreign and domestic companies. By indirectly increasing state interference with markets also in the exporting countries, policies may effectively bring about what they are supposed to fight: public protection of a dumper's home market.

Antidumping policies are also closely intertwined with other trade-protective measures such as export-restraint arrangements, which impair economic welfare as well. Actually, the two forms of intervention have proved

---

[88] OECD, *Competition Policy and Antidumping: Summary and Conclusions* [mimeo] Paris, October 1995, p. 7.

[89] U.S. International Trade Commission, *The Economic Effects of Antidumping and Countervailing Duty Orders and Suspension Agreements*, Washington, D.C., June 1995.

to be effective complements, as many antidumping cases were superseded by "voluntary" export restraints (VER). The threat of formal action under the antidumping law was the lever to force an exporter to accept a VER. It has also been shown, e.g., in the U.S.-Japan DRAMs case, that antidumping measures originally targeted at a limited number of product variants finally extended into an agreement covering the whole range of the respective product group.[90]

Dumping and Antidumping in High Technology

In high technology, the two basic forms of "dumping," i.e., regional price differentiation and (temporary) pricing below cost, seem to be widespread business practices reflecting central characteristics of the sector such as market segmentation, high fixed costs (especially up-front R&D expenditures), decreasing variable costs over time due to learning by doing, and first-mover advantages. All this should make high-technology producers particularly vulnerable to antidumping measures, and especially so in the early stages of production.

Competition in high-technology markets has also been described as demand-driven product differentiation, where the ability of suppliers to accurately forecast consumers' tastes (thereby creating temporary market access barriers subject to quick erosion) counts more than lasting entry deterrence based on an exclusive mastery of the technology at hand.[91] This proposition is consistent with the empirical finding of relatively low degrees of supplier concentration in many high-technology markets in the exporting countries, pointing to little leeway for predatory pricing. On balance, there are even more possibilities—and stronger incentives—for a protectionist use of the antidumping instrument in high technology than in "ordinary" business. In many cases, calculated dumping margins, and actual duties imposed, are indeed extraordinarily high for high-technology products.[92] Even though, in terms of numbers of cases, high-technology products have

---

[90] In fact, the DRAMs under antidumping measures represented only 7 percent of the whole semiconductor sector. See P.A. Messerlin, *Reforming the Rules of Antidumping Policies*. Paper presented at the conference "Towards a New Global Framework for High-Technology Competition," Kiel, Germany, 30–31 August 1995, pp. 7–8.

[91] Messerlin points to the general availability of basic technologies, which is in most cases also suggested by the fact that many complainants in U.S. and EC antidumping cases involving high-technology products were the technological leaders of the product a couple of years before lodging the complaints. See P.A. Messerlin, *Reforming the Rules of Antidumping Policies*, p. 4.

[92] For instance, in the U.S. EPROMs case, the calculated dumping margins ranged from 60 to 188 percent, in the EC DRAMs case from 8 to 378 per cent. See P.A. Messerlin, *Reforming the Rules of Antidumping Policies*, p. 9.

not been the main target of antidumping policies, their incidence is high if measured by product (wide coverage of affected product groups), market (high volumes of sales and trade), and supplier (broad representation of major producers) characteristics. The largest case in the whole EC anti-dumping history, for instance, concerns plain-paper copiers, with imports of US $3.5 billion in 1992. In this case, 20 percent duties on Japanese photo-copiers, originally introduced in 1987, were reimposed (and extended to larger models) in October 1995, for three more years. The decision was reached by the Council of Ministers with a close (8–7), majority which was sufficient only after voting rules were changed (from qualified to simple majority) in March 1994. European imports of Japanese photocopiers have dropped sharply since the introduction of antidumping duties, but the market share of Japanese producers has nonetheless increased, as many of them have moved production inside the European Union.[93]

Empirical analysis for a number of electronic products (color television sets, compact disc players, and plain paper photocopiers) from Japan on which antidumping duties were imposed by the United States and the European Community reveals a strong tendency of antidumping policies in this sector to protect domestic competitors rather than preserve competition in the long run.[94] The case studies investigate certain structural characteristics of the respective markets and industries—such as entry barriers, relative home-market size, concentration ratios, market shares, and static and dynamic economies of scale—that are potentially conducive to predatory or strategic dumping.[95]

Three results of the studies stand out:

First, declining government protection by tariff and non-tariff barriers as well as subsidies contrasts with persisting low import penetration of the examined Japanese markets. Available evidence on private restrictions against import competition—such as incompatible product standards agreed by domestic companies, entry-deterring horizontal collusion among suppliers, and vertical restraints through exclusive distribution channels—is unable to ex-

---

[93] See *Financial Times,* 19 October 1995 (G. de Jonquières and E. Tucker, "Pressure Grows for EU to Overhaul Dumping Policy") and 15 September 1995 (E. Tucker, "EU Split over Dumping Duties on Japanese Copiers").

[94] See P.A. Messerlin and Y. Noguchi, *Antidumping Policies and International Trade of Electronic Products* [mimeo], Paris, OECD, August 1995.

[95] Predatory dumping refers to low-priced exporting with the intention of driving rivals out of business in order to obtain monopoly power in the importing market, whereas strategic dumping describes exporting that injures domestic rivals through an overall strategy or general circumstances of the exporting nation, encompassing both the pricing of the exports and restraints protecting the exporter's home market. See R.D. Willig, *The Economic Effects of Antidumping Policy* [mimeo], Paris, OECD, October 1995, p. 6f.

plain the phenomenon. In fact, low degrees of supplier concentration in the respective markets in Japan indicate rather that private entry barriers are moderate. Moreover, a limited number of domestic newcomers and foreign investors have successfully entered the markets considered in the analysis, the existence of interlocking "keiretsu" systems notwithstanding.

Second, factual exclusion of American and European exporters from the investigated Japanese markets has not evidently disadvantaged these companies in international competition. In particular, it has not prevented them from exploiting scale economies in production, nor discouraged R&D efforts which for their profitability hinge on the number of units produced. Reliance on OEM (Original Equipment Manufacturing) agreements, for instance, by which firms are allowed to sell products of other companies under their own brand name, permits segment-specific economies of scale to be utilized without constraining the product range of each supplier.

Finally, the case studies demonstrate that the recoupment of initial losses (and harvesting of some additional gain) on which the success of any predatory scheme crucially depends, is unlikely to happen. An important factor here is (actual or potential) market entry by third-country suppliers, e.g., European firms in the United States, challenging possible dominant positions of potential predators. It has also been shown that access to the basic technologies needed is relatively easy, which increases the contestability of the markets in question.

In sum, little evidence on the possible existence of predatory or strategic dumping practices has been found in empirical analysis.

The Uruguay Round Approach to Antidumping

The degeneration of antidumping measures into protective, selective, anticompetitive, and strategic trade policy devices has not been effectively contained in the Uruguay Round. The agreed changes will hardly transform antidumping action into competition policy. An attempt was made to correct some of the biases embodied in national anti-dumping regulations and to remove inconsistencies in the use of procedures where rules often changed between cases and even methods within cases. However, no standards of competition law were adopted, and a number of highly questionable practices were allowed to continue as long as they are "appropriately" explained. Of particular relevance for high-technology products in this context are the modifications regarding averaging methods in price comparisons and the treatment of start-up costs.

In certain circumstances, and if appropriately explained, the new antidumping agreement allows the national authorities to deviate from the general method of comparing average home prices with average export prices, or individual domestic transactions with corresponding individual transactions abroad. Instead, dumping investigators may compare a weighted *average* of domestic prices (omitting, if necessary, low-priced sales outside the "ordinary course of trade") with *individual* export prices if the latter vary significantly among different purchasers, regions, or time periods. As this allows high-priced exports to be disregarded in the calculations, it is a sure way of finding dumping in almost every case, and particularly in cases where prices (and models) are as diverse and change as quickly as in high-technology products.

It may be an easy task for policymakers to offer plausible explanations for the chosen calculation method, which withstands any possible challenge by the affected trading partners with relative ease. Past dispute settlement panels in antidumping cases in general used to treat lack of explanation as reason to recommend that an antidumping duty be revised, not as reason for it to be removed. With the new general ("integrated") dispute settlement mechanism in place, this could possibly change in the future, as panel decisions no longer need the consent of the defending country. However, the specific dispute settlement section of the antidumping agreement gives the national determination of dumping margins the benefit of the doubt. It is explicitly recognized that the provisions of the agreement admit of more than one permissible interpretation. The panel is only to ask whether the antidumping authorities reached their conclusion without explicit error of fact or reasoning. This will make it very difficult for it to reject an explanation that has even the slightest plausibility.[96]

The provisions of the agreement concerning start-up costs or sales below total unit costs (including a "reasonable" profit margin) are equally ambiguous. On the one hand, below-cost sales in the exporter's market have to be eliminated in the determination of the "normal value" if certain conditions are given.[97] This *raises* the "normal value" and so tends to make the finding of dumping even more likely than before (when elimination was a mere possibility). On the other hand, the agreement for the first time requires nonrecurring items of cost (like R&D expenditures), which benefit future production, to be allocated over a longer period of time. In the case of start-up operations the costs at the end of the start-up period are relevant. This should *lower* the calculated normal value and hence *constrain* the finding of dumping. It could be particularly important in cases of low home-country sales (less than 5 percent of the disputed export sales) where

---

[96] See B. Hindley, "Two Cheers for the Uruguay Round," in: *Trade Policy Review 1994,* London, September 1994, p. 27.

[97] Below-cost sales must occur over an extended period of time (normally one year), in substantial quantities (at least 20 percent of the total sales under consideration), and "at prices which do not provide for the recovery of all costs within a reasonable period of time" (Article 2 of the Agreement on Implementation of Article VI of GATT 1994).

constructed values instead of direct price comparisons are used in the determination of dumping margins.

Other modifications of antidumping rules agreed in the Uruguay Round
are a mixed blessing as well. The five-year sunset clause, according to
which antidumping measures automatically expire after five years, was adapted
from existing EU legislation. However, its value appears to be limited: (1)
not only is it possible to extend the limit (and this is actually happening, as
recently shown in the EU-Japan copier case noted above) but (2) the limit
has also apparently not prevented European antidumping enforcement from
being as biased as U.S. practices. For similar reasons, the new de minimis
provision—which requires the immediate termination of an antidumping
investigation when the margin of dumping is less than 2 percent of the
export price, or the volume of dumped imports less than 3 percent of the
respective total imports—seems unlikely to contain harassment of exporters
in an effective way.

At the same time, the more elaborate requirements for an antidumping
investigation to be initiated after application may effectively lengthen the
period of uncertainty about trading conditions, and thus discourage imports,
since there is no time limit in the pre-initiation phase as compared with the
subsequent stages. With regard to the determination of injury and the causal
link between dumping and injury, the new agreement provides little change
from the preceding antidumping code of the Tokyo Round, and so allows
considerable scope for arbitrary decisionmaking (e.g., in the definition of
"injured" domestic industries, which may *in*clude or *ex*clude foreign subsidiaries) to continue.

By and large, the substantive and procedural conditions imposed by the
new antidumping agreement on policies in this area appear unable in themselves to effect substantial change. Some of the improvements reached
have also been eroded in national implementation legislation reflecting the
political power of protection-seeking interests. Apparently, even the small-
est content of ambiguity of any word has been intensely exploited. Amend-
ing antidumping procedures has been depicted as an activity with a productivity close to Sisyphus's; new provisions seem to be deviated from their
initial purpose at a more rapid pace than they were introduced.[98] However,
antidumping rules and practices remain far removed from the standards of
sound competition policy which they were originally supposed to supplement at the international level. Disciplining antidumping policies should
therefore rank high among the market access issues on the post-Uruguay
Round trade agenda.

---

[98] See P.A. Messerlin, *Reforming the Rules of Antidumping Policies.* Paper presented at
the conference "Towards a New Global Framework for High-Technology Competition," Kiel,
Germany, 30–31 August 1995, p. 12.

Reforming Antidumping Policy

The first-best solution, from an intervention-theoretic point of view, would be to substitute competition rules and procedures, either harmonized and coordinated among nation-states or truly globalized, for existing antidumping laws. This would allow the removal of distortions, such as abuse of a dominant market position in the exporting country, at the source, rather than merely compensating for them. However, this is a long-term perspective at best, unlikely to materialize in the near future. It sharply contrasts with political realities, which are characterized by a proliferation of antidumping regulations and an unwillingness of the most prominent user countries and regions (the United States and the European Union) to dispense with an expedient instrument of trade policy. Protection-seeking interests would have to be accommodated in one form or another anyway, which would leave little net value in abolishing antidumping measures.

The prevailing political view is of raising international standards in the area of competition law, thus reducing the need to invoke the antidumping law, i.e., competition laws working effectively alongside the antidumping law without substituting for it. In fact, there have been only a few cases of anti-dumping policies being superseded by competition policies. Apart from the European Union, these are the formation of the European Economic Area and, to a lesser extent, the Australian-New Zealand Closer Economic Relations Trade Agreement.

In the face of the widely perceived political need to retain the antidumping option in trade policy, reforming policy in this area could proceed on four tracks:

- Removal of trade and investment barriers in the exporting (and importing) countries
- Ad-hoc review of the state of competition (actual and potential) in the exporting-country markets and of existing public barriers to entry (and presence) by foreign competitors
- "Importing" competition-policy standards into antidumping legislation, and raising the legal standing of consumer and industrial-user interests in the investigations
- Developing international competition rules.

The *first* track involves ongoing liberalization of trade on a reciprocal basis, including the removal of remaining tariff, non-tariff, and regulatory barriers to market access raised by governments, as well as the negotiation of an international investment agreement that guarantees foreign companies freedom of establishment and national treatment in host countries. This will significantly change the environment for predatory behavior and make it even less likely to happen. By the same token, it weakens the legitimation bases of antidumping policies. However, as international negotiations on broadly based liberalization and deregulation take time, no quick impact on dumping practices and antidumping measures can be expected.

A more immediate impact would have a multilateral agreement, as proposed by Hoekman and Mavroidis (1994),[99] providing for an examination (preferably by competition authorities) of market characteristics (including private and public entry barriers) in the exporting country to precede any antidumping investigation. The policy objective in this *second* track would be to establish whether the structural preconditions for economically harmful (i.e., reducing the importing country's welfare) dumping strategies are given or not. An antidumping petition would consequently have to be turned down from the outset, irrespective of the existence of dumping, if (1) the competition agency in the exporting country should find the accused firms neither engaged in anticompetitive practices nor benefiting from government-created or supported market access barriers against foreign companies, and (2) the corresponding authorities of the importing country agree with the finding.

In the *third* track competition aspects are considered after an antidumping investigation has been launched. The focus now shifts from the exporting to the importing country market, where the quality of the injury inflicted on domestic industries through foreign dumping apparently varies depending on the state of competition. In highly concentrated markets with high entry barriers, for instance, "injury" often means a reduction of noncompetitive excess returns. "Dumping" would correspondingly work to "disrupt" monopolistic market structures—to the benefit of other domestic agents—and to enhance general welfare.[100] The injury test in antidumping laws should accordingly be redesigned and expanded so as to more comprehensively include the interests of consumers and industrial users and review the state of competition in the domestic markets concerned. It could possibly also include an examination of whether antidumping duties would help domestic industries to compete more effectively.

A more competition-oriented approach to antidumping cases could involve the adoption of competition-policy concepts such as "relevant mar-

---

[99] See B.M. Hoekman, and P.C. Mavroidis, *Antitrust-Based Remedies and Dumping in International Trade.*

[100] For example, Tharakan finds most of the EU antidumping impositions to have taken place in favor of industries which have a high degree of concentration in the European Union. He concludes that the lobbying power of oligopolistic industries is effectively used by them to obtain protection from import competition through antidumping measures. See P.K.M. Tharakan, "AntiDumping Policy and Practice of the European Union: An Overview," *Economisch en Sociaal Tijdschrift*, 1994, no. 4, p. 565.

ket" and "critical degrees of concentration" in antidumping investigations, in order to correct the biases inherent in current antidumping concepts such as "like product" and "major proportion of the domestic industry." Defining quantified "thresholds" (comparable to the "tariffication" process in the Uruguay Round) would facilitate multilateral negotiations in this area. In addition to these reinterpretations of existing antidumping provisions, the appeals system in antidumping cases could be substantially improved. As proposed by Messerlin,[101] an antidumping case could have to be passed from antidumping authorities to antitrust agencies, e.g., from DG I to DG IV in the European Commission or to the cartel offices in individual EU member states, should a court ruling repeal the case for the formers' neglect of competition aspects. In sum, to the three elements of an antidumping investigation (unfair trade, injury, and causal link between the two) a fourth element—competition analysis—would be added.[102] It would be a logical complement to the pre-test of competition on the exporting-country market noted above.

The *fourth* track entails the codification of international competition rules. This would contribute importantly to an eventual elimination of antidumping policies. Their final replacement by competition policies is feasible, though, only with a high standard of integration between countries. This would require realization of the "four freedoms" (i.e., complete liberalization of trade in goods, services, labor, and capital) plus agreement on common disciplines for, or mutual recognition of, certain policies and regulations regarding in particular subsidies, government procurement, technical standardization, services, and for that matter competition. Removing the antidumping instrument accordingly seems to be possible in the foreseeable future only for limited numbers of countries and typically in a regional context. This, again, bears chances as well as risks, since increased competition—and related adjustment pressures—inside the "club" might provoke compensatory antidumping action against outsiders.

Toward Multilateral Competition Rules

International coordination, or harmonization, of competition policies may not only help to contain dumping practices and counter-productive antidumping policies, but also prevent anticompetitive business conduct in international trade in general. This may essentially take three forms:

---

[101] See P.A. Messerlin, *Reforming the Rules of Antidumping Policies*, p. 13.

[102] See also H. Vandenbussche, "How Can Japanese and Central European Exporters to the European Union Avoid Antidumping Duties?" *World Competition*, March 1995.

- Exclusion of foreign suppliers from domestic markets by the exercise of horizontal market power through collusive behavior or vertical restraints of competition such as exclusive dealerships between companies of one country. This may effectively—and asymmetrically—thwart the liberalization of trade negotiated by governments.

- Formation of export cartels among a country's firms, and similar action to inhibit foreign competitors in international markets, with the ultimate effect of significantly raising international prices. With asymmetric incidence of private trade barriers between countries, trading partners may in this case even suffer net losses from "public" liberalization.

- Restraint of competition between companies from different countries in the form of strategic alliances (frequently of a technological nature), cartels, mergers, and aquisitions in particular. Again, this could impair the benefits of liberalization. However, in the majority of cases, international cooperation among firms seems to stimulate rather than reduce competition on national markets, as these become increasingly global in the course of trade liberalization.

On this background, international policymaking in the field of competition faces two major challenges:

- Adjusting existing rules and procedures in national competition policies—or building them from scratch—and monitoring their implementation, so as to remove—and prevent—trade-restrictive and mercantilistic biases in competition policy. In particular, strict enforcement of the agreed provisions would have to be guaranteed and export cartels prohibited if they go beyond a mere sharing of expertise and services in supplying foreign markets. Export cartels may also be conducive to collusive behavior on domestic markets.

Devising principles—and practical guidelines—for the international cooperation among national agencies responsible for competition policy, in order to resolve conflicts that may arise when private restraints of competition simultaneously affect a number of national markets. Such conflicts are likely to occur also if competition laws—and the weight given to competition as compared with other policy aims—are largely identical between the countries concerned. This is because in many cases the impact of the anticompetitive practices will vary from country to country.[103] In the final analysis, international action may require a supranational authority to coor-

---

[103] With markets becoming increasingly global, international antitrust frictions of this kind should, however, tend to diminish. See R. Jungnickel and G. Koopmann, "Globalization of Business—Implications for International Competition and Related Policies," in E. Kantzenbach, H.-E. Scharrer, and L. Waverman (eds.), *Competition Policy in an Interdependent World Economy,*. Hamburg, 1993, pp. 44–45.

dinate the collection of information, and to apply any sanctions to the firm or firms involved.[104]

In general, greater market interpenetration due to the integration of national markets has resulted in the need for competition policies to adjust to different circumstances of international competition.[105] More specifically, and with particular relevance to high technology, an important rationale, denoted by Soete,[106] for international competition policy aimed at counteracting the emergence of worldwide cartels between global firms could be to challenge national strategic policies based on arguments of dependence on "foreign" monopoly pricing. Other reasons for specific (national as well as international) antitrust concerns in high technology refer to (1) *network externalities* (e.g., in computer software), which could bias industries toward monopoly, as the value placed on network membership by a consumer rises with the number of other people on the network; (2) *systems leverage,* by which a firm which controls one part of a system may spread its monopoly into others through, e.g., cross-subsidization;[107] (3) *standardization,* which may lend individual firms, and groups of companies, considerable market power, as consumers and user firms get "locked" into the system and as entry barriers, through costs of "switching" systems, are raised; and (4) *innovation cartels,* inhibiting the innovation process proper (by reducing the number of innovative efforts) and possibly "spilling over" into other fields of activities such as production and marketing.[108]

The actual development of international competition rules is still in its infancy. It could be advanced through "case law" built in dispute settlement procedures based on existing GATT provisions. Of particular interest in this context is the *non-violation* clause of GATT Article XXIII:1(b), designed to address the concern of GATT members relating to a modification of negotiated concessions through subsequent government action in

---

[104] See D.A. Hay, *The Economic and Trade Effects of Anti-Competitive Practices in a Globalising World Economy. [mimeo],* Paris, OECD, 26 April 1995.

[105] See P. Lloyd and G. Sampson, "Competition and Trade Policy: Identifying the Issues after the Uruguay Round," *The World Economy,* vol. 18, no. 5, September 1995, p. 686.

[106] See L.G. Soete, *Technology Policy and the International Trading System: Where Do We Stand?* Paper presented at the conference "Towards a New Global Framework for High-Technology Competition," Kiel, Germany, 30–31 August 1995, p.15.

[107] For instance, Microsoft, the leading computer-software firm, has been accused of using profits from the operating-system market to subsidize applications programs. See "Thoroughly Modern Monopoly," *The Economist,* 8 July 1995, p. 92.

[108] However, empirical evidence on competitive restraints involved in, and flowing from, technological cooperation is scarce. A possible loss of variety seems to be more than compensated for by substantial economies of scale and gains in efficiency. For an overview of trends in (national and international) technological partnering see J. Hagedoorn, *The Economics of Cooperation Among High-Tech Firms—Trends and Patterns in Strategic Technology Partnering since the Early Seventies,* in: G. Koopmann and H.E. Scharrer (eds.), *The Economics of High-Technology Competition and Cooperation in Global Markets,* Baden-Baden, 1996, p. 173ff .

areas that either were not addressed by the GATT or did not violate a GATT obligation. An example could be an exemption by the competent antitrust authorities granted to private enterprises, that effectively reduces market access opportunities for products of third countries by establishing difficult-to-penetrate distribution channels.[109] In conjunction with a significantly strengthened dispute settlement procedure, nonviolation complaints could therefore lead to the opening of foreign markets closed by private restrictive practices. However, the wording of the clause is too general, and its reach too limited, for it to serve as a solid base for an effective international competition policy. Its "structural" weakness is to be, by definition, not based on explicit provisions concerning anticompetitive business conduct in international trade (with the exception of dumping).

This is different with the *Draft International Antitrust Code (DIAC)* proposed by an international group of legal experts.[110] The Draft Code provides for minimum standards to be observed by the antitrust authorities in the participating countries (which need not include all WTO members, as the DIAC is designed as a *plurilateral* GATT-WTO agreement) in dealing with private restraints of competition that affect international trade flows. The practices covered range from the formation of international cartels to the abuse of dominant market positions and from horizontal to vertical restraint. An international competition agency, equipped with the right of an *International Procedural Initiative*, would be entrusted to safeguard the application of the national law in cases where an inactive member state would not take its own initiative; if necessary, it could sue in the national courts to ensure application. Conflicts arising from impacts in one country of competitive restraints originating in another would be addressed by taking advantage of the new WTO dispute settlement procedure. Even though this proposal as a whole is fraught with a number of difficulties regarding its practical implementation and enforcement, as well as inevitable losses of national sovereignty,[111] it seems in principle to be well suited to meet the challenges noted above, including the specific antitrust issues related to high technology. It could effectively flank the international trading order with an international competition regime.

---

[109] See B.M. Hoekman, and P.C. Mavroidis, *Antitrust-Based Remedies and Dumping in International Trade*, Policy Research Working Paper 1347, The World Bank, Washington, D.C., August 1994, p. 20. A comprehensive discussion of non-violation in the competition context is given in B.M. Hoekman, and P.C. Mavroidis, "Competition, Competition Policy and the GATT," *The World Economy*, vol. 17, 1994, pp. 121–150.

[110] For the text and a detailed explication see W. Fikentscher, "Competition Rules for Private Agents in the GATT/WTO System," *Außenwirtschaft*, vol. 49, no. 2/3 1994, pp. 281–325.

[111] Klodt, regarding the proposed *minimum* standards rather than *maximum* standards, points to the particular problems of establishing an international merger control, as well as of rules for the control of an abuse of market power. See H. Klodt, "Internationale Regeln für den Wettbewerb," *Wirtschaftsdienst*, October 1995, p. 560.

References

Aaron, D.L., "After the GATT, U.S. Pushes Direct Investment," *The Wall Street Journal*, 2 February 1995.

Aghion, P., and P. Howitt, "A Model of Growth through Creative Destruction," *Econometrica*, vol. 60 (1992), pp. 323-351.

Baldwin, R., "The Economics of Technological Progress and Endogenous Growth in Open Economies," in G. Koopmann and H.-E. Scharrer (eds.), *The Economics of High-Technology Competition and Cooperation in Global Markets*. Baden-Baden, 1996, p. 63–75.

Baldwin, R., and H. Flam, "Strategic Trade Policies in the Market for 30-40 Seat Commuter Aircraft," *Weltwirtschaftliches Archiv/Review of World Economics*, vol. 125 (1989), pp. 484–500.

Baldwin, R., and P.R. Krugman, "Industrial Policy and International Competition in Wide-Bodied Jet Aircraft," in R. Baldwin (ed.), *Trade Policy Issues and Empirical Analysis*. Chicago, 1988.

Beekmann, H., "The 1994 Revised Commercial Policy Instrument of the European Union," *World Competition*, vol. 19 (1995) no 1., pp. 53–75.

Bletschacher, G., and H. Klodt, *Strategische Handels-und Industriepolitik*. Kieler Studien No. 244. Tübingen, 1992.

Brander, J.A., and B.J. Spencer, "Export Subsidies and International Market Share Rivalry," *Journal of International Economics*, vol. 16 (1985), pp. 83–100.

Brander, J.A., and B.J. Spencer, "International R&D Rivalry and Industrial Strategy," *Review of Economic Studies*, vol. 50 (1983), pp. 707–722.

Bresnahan, T.F., and M. Trajtenberg, *General Purpose Technologies: "Engines of Growth"?* NBER Working Paper No. 4148, National Bureau of Economic Research, Cambridge, Mass., August 1992.

Caballero, R.J., and R.K. Lyons, "The Case for External Effects," in A.Cukierman, (ed.), *Political Economy, Growth, and Business Cycles*. Cambridge, Mass., 1992, pp. 117–139.

Coe, D.T., and E. Helpman, *International R&D Spillovers*. NBER Working Paper No. 4444. National Bureau of Economic Research, Cambridge, Mass., August 1993.

Coe, D.T., E. Helpman, and A.W. Hoffmaister, *North-South R&D Spillovers*. CEPR Discussion Paper No. 1133. Centre for Economic Policy Research, London, February 1995.

Cohen, S., and P.H. Chin, *Tipping the Balance: Trade Conflicts and the Necessity of Managed Competition in Strategic Industries*. Paper presented at the conference "Towards a New Global Framework for High-Technology Competition," Kiel, Germany, 30–31 August 1995.

Cordon, M., *Trade Policy and Economic Welfare*. Oxford, 1974.

Department of Foreign Affairs and International Trade, *Register of United States Barriers to Trade, 1994*. Ottawa, 1994.

Dhar, B., "The Decline of Free Trade and U.S. Trade Policy Today," *Journal of World Trade,* vol. 26 (1992) no. 6, pp. 133–154, based on: Commission of the European Communities, *Report on United States Trade Barriers and Unfair Trade Practices,* Brussels, 1991.

Dixit, A.K., and A.S. Kyle, "The Use of Protection and Subsidies for Entry Promotion and Deterrence," *American Economic Review,* vol. 75 (1985), pp. 139–152.

Economist, Uncle Sam's Helping Hand," pp. 91–93, 2 April 1994.

Economist, "Scrempp Cocktail," p. 70, 20 May 1995.

Economist, "Thoroughly Modern Monopoly," p. 92, 8 July 1995.

Economist, "A Tale of Two Conglomerates," p. 20, 18 November 1995.

Economist, "Dismantling Daimler," pp.79-80, 18 November 1995.

Fikentscher, W., "Competition Rules for Private Agents in the GATT/WTO System," *Außenwirtschaft,* vol. 49, no. 2/3, 1994, pp. 281–325.

Financial Times, "Attack on Anti-Dumping Law Sparks OECK Row" (G. de Jonquieres), 21 September 1995.

Financial Times, "Pressure Grows for EU to Overhaul Dumping Policy" (G. de Jonquieres and E. Tucker), 19 October 1995.

Financial Times, "EU Split over Dumping Duties on Japanese Copiers" (E. Tucker), 15 September 1995.

Finger, J. Michael, *Legalized Backsliding: Safeguard Provisions in the GATT.* [Mimeo.] Washington, D.C., 26–27 January 1995.

GATT, *News of the Uruguay Round,* April 1994.

Glaeser, E.L., H.D. Kallal, A. Scheinkman, and A. Shleifer, "Growth in Cities," *Journal of Political Economy,* vol. 100 (1992), pp. 1126–1152.

Graham, E.M., "Investment and the New Multilateral Trade Context," in OECD, *Market Access After the Uruguay Round: Investment, Competition and Technology Perspectives.* Paris, 1996, pp. 35–62.

Griliches, Z., "The Search for R&D Spillovers," *Scandinavian Journal of Economics,* vol. 94 (1992), Supplement, pp. 24–47.

Grossman, G.M., "Promoting New Industrial Activities: A Survey of Recent Arguments and Evidence," *OECD Economic Studies,* vol. 14 (1990), pp. 87–125.

Grossman, G.M., and E. Helpman, *Innovation and Growth in the Global Economy.* Cambridge, Mass., 1991.

Hagedoorn, J., "The Economics of Cooperation Among High-Tech Firms— Trends and Patterns in Strategic Technology Partnering since the Early Seventies," in G. Koopmann, and H.-E. Scharrer (eds.), *The Economics of High-Technology Competition and Cooperation in Global Markets.* Baden-Baden, 1996. Pp.173–198.

Hay, D.A., *The Economic and Trade Effects of Anti-Competitive Practices in a Globalising World Economy.* [Mimeo] OECD, Paris, 26 April 1995.

Helpman, E., and M. Trajtenberg, *A Time to Sow and a Time to Reap: Growth Based on General Purpose Technologies*. NBER Working Paper No. 4854. National Bureau of Economic Research, Cambridge, Mass., September 1994.

Hindley, B., "Two Cheers for the Uruguay Round," in *Trade Policy Review 1994*. London, September 1994.

Hoekman, B.M., and P.C. Mavroidis, "Competition, Competition Policy and the GATT," *The World Economy*, vol. 17 (1994) pp. 121–150.

Hoekman, B.M., and P.C. Mavroidis, *The WTO's Agreement on Government Procurement: Expanding Disciplines, Declining Membership?* Discussion Paper 1112. CEPR, London, 1995.

Hoekman, B.M., and P.C. Mavroidis, *International Antitrust Policies for High-Tech Industries?* Paper presented at the conference "Towards a New Global Framework for High-Technology Competition," Kiel, Germany, 30–31 August 1995. Pp. 10f.

Hoekman, B.M., and P.C. Mavroidis, *Antitrust-Based Remedies and Dumping in International Trade*. Policy Research Working Paper 1347. The World Bank, Washington, D.C., August 1994.

Irwin, D. A., *Trade Politics and the Semiconductor Industry*. NBER Working Paper No. 4745. National Bureau of Economic Research, Cambridge, Mass., May 1994.

Irwin, D.A., and P.J. Klenow (1994a), "Learning-by-Doing Spillovers in the Semiconductor Industry," *Journal of Political Economy*, vol. 102 (1994), pp. 1200–1227.

Irwin, D.A., and P.J. Klenow (1994b), *High Tech R&D Subsidies: Estimating the Effects of SEMATECH*. NBER Working Paper No. 4974. National Bureau of Economic Research, Cambridge, Mass., December 1994.

Jaffe, A.B., M. Trajtenberg, and R. Henderson, "Geographic Localization of Knowledge Spillovers as Evidenced by Patent Citations," *Quarterly Journal of Economics*, vol. 108 (1993), pp. 577–598.

Jungnickel, R., and G. Koopmann, "Globalization of Business—Implications for International Competition and Related Policies," in E. Kantzenbach, H.-E. Scharrer, and L. Waverman (eds.), *Competition Policy in an Interdependent World Economy*. Hamburg, 1993. Pp. 33–50.

Kantzenbach, E., H.-E. Scharrer, L. Waverman (eds.), *Competition Policy in an Interdependent World Economy*. Hamburg, 1993.

Kleinfeld, G., "Taxation of Automobiles and GATT National Treatment Obligations: Where to Draw the Line?" *World Competition*, vol. 19 (1995) no. 1, pp. 77–90.

Klepper, G., "Entry into the Market for Large Transport Aircraft," *European Economic Review*, vol. 34 (1990), pp. 775–803.

Klepper, G., "Industrial Policy in the Transport Aircraft Industry," in P. Krugman and A. Smith (eds.), *Empirical Studies of Strategic Trade Policy*. Chicago and London, 1994. Pp. 101–126.

Klodt, H., et al., *Forschungspolitik unter EG-Kontrolle.* Kieler Studien No. 220. Tübingen, 1988.

Klodt, H., J. Stehn et al., *Strukturpolitik der EG.* Kieler Studien No. 249. Tübingen, 1992.

Klodt, H., *Wettlauf um die Zukunft.* Kieler Studien No. 206. Tübingen, 1988.

Klodt, H., "Internationale Regeln für den Wettbewerb," *Wirtschaftsdienst,* October 1995.

Koopmann, G., and H.-E. Scharrer (eds.), *The Economics of High-Technology Competition and Cooperation in Global Markets.* Baden-Baden, 1996.

Krugman, P., and M. Obstfeldt, *International Economics: Theory and Policy,* 3rd edition. New York, 1994.

Laffont, J.-J., and J. Tirole, "Auction Design and Favoritism," *International Journal of Industrial Organization,* vol. 9, 1991, no. 1, pp. 9–42.

Lawrence, R.Z., "Towards Globally Contestable Markets," in OECD, *Market Access after the Uruguay Round: Investment, Competition and Technology Perspectives.* Paris, 1996. Pp. 25–33.

Lloyd, P., and G. Sampson, "Competition and Trade Policy: Identifying the Issues after the Uruguay Round, *The World Economy,* vol. 18, no. 5, September 1995.

Maitland, I., "Who Won the Industrial Policy Debate?" *Business & the Contemporary World,* 1, 1995, pp. 83–95.

Messerlin, P.A., "Agreement on Public Procurement," in OECD, *The New World Trading System:* Reading, OECD Documents, Paris, 1994. Pp. 65–71.

Messerlin, P.A., *Reforming the Rules of Antidumping Policies.* Paper presented at the conference "Towards a New Global Framework for High-Technology Competition," Kiel, Germany, 30–31 August 1995.

Messerlin, P.A., and Y. Noguchi, *Antidumping Policies and International Trade of Electronic Products,* [Mimeo.] OECD, Paris, 30 August 1995.

Nicolas, F., and J. Repussard, *Common Standards for Enterprises.* Luxembourg, 1988.

OECD, *Trade and Competition Policies: Comparing Objectives and Methods.* Trade Policy Issues, no. 4. Paris, 1994.

OECD, *Competition Policy and Antidumping, Summary and Conclusions.* [Mimeo] Paris, October 1995.

Office of Technology Assessment, *Global Standards: Building Blocks for the Future.* U.S. Congress, Washington, D.C., 1992.

Ostry, S., "Technology Issues in the International Trading System," in OECD, *Market Access after the Uruguay Round: Investment, Competition and Technology Perspectives.* Paris, 1996. Pp. 145–170.

Ostry, S., and R.R. Nelson, *Techno-Nationalism and Techno-Globalism: Conflict and Cooperation*. Brookings Institution, Washington, D.C., 1994.

Paqué, K.-H., "Technologie, Wissen und Wirtschaftspolitik—Zur Rolle des Staates in Theorien des endogenen Wachstums," in *Die Weltwirtschaft* 3 (1995), pp. 237–253.

Rivera-Batiz, F.L., "The Economics of Technological Progress and Endogenous Growth in Open Economies," in G. Koopmann, and H.-E. Scharrer (eds.), *The Economics of High-Technology Competition and Cooperation in Global Markets*. Baden-Baden, 1996. Pp. 31–62.

Rivera-Batiz, F.L., and P.M. Romer, "International Trade with Endogenous Technological Change," *European Economic Review*, vol. 35 (1991a), pp. 971–1004.

Rivera-Batiz, F.L., and P.M. Romer, "Economic Integration and Endogenous Growth," *Quarterly Journal of Economics*, vol. 106 (1991b), pp. 531–556.

Schott, J. (assisted by J.W. Buurman), *The Uruguay Round: An Assessment*. Institute for International Economics, Washington, D.C., November 1994.

Schott, J.J., *Dispute Settlement in the Multilateral Trading System and High-Technology Trade*. Paper presented at the conference, "Towards a New Global Framework for High-Technology Competition," Kiel Institute of World Economics, Kiel, Germany, 30–31 August 1995.

Soete, L., *Technology Policy and the International Trading System: Where Do We Stand?* Paper presented at the conference "Towards a New Global Framework for High-Technology Competition," Kiel, Germany, 30–31 August 1995.

Stehn, J., "Wettbewerbsverfälschungen im Binnenmarkt: Ungelöste Probleme nach Maastricht," in *Die Weltwirtschaft,* 1993, pp. 43–60.

Stehn, J., *Subsidies, Countervailing Duties, and the WTO: Towards an Open Subsidy Club*. Kiel Discussion Paper. Kiel Institute of World Economics, Kiel, June 1996.

Stern, R., and B.M. Hoekman, "The Codes Approach," in J.M. Finger and A. Olechowski (eds.), *The Uruguay Round: A Handbook on the Multilateral Trade Negotiations*. The World Bank, Washington, D.C., 1987.

Stolpe, M., "Industriepolitik aus Sicht der neuen Wachstumstheorie," in: *Die Weltwirtschaft*, 1993, no. 3, pp. 361–377.

Stolpe, M., *Technology and the Dynamics of Specialization in Open Economies*. Kiel Studies 271. Tübingen, 1995.

Tharakan, P.K.M., "Anti-Dumping Policy and Practice of the European Union: An Overview," in *Economisch en Sociaal Tijdschrift*, 1994, No. 4.

Tyson, L.D., *Who's Bashing Whom? Trade Conflict in High-Technology Industries*. Washington, D. C., 1992.

UNCTAD, *Trade and Development Report, 1994*. United Nations, 1994. P. 137.

U.S. International Trade Commission, *The Economic Effects of Antidumping and Countervailing Duty Orders and Suspension Agreements.* Washington, D.C., June 1995.

USIS, "U.S. Government Fact Sheet: U.S.-Japan Auto and Auto Parts Agreements," *U.S. Information and Texts,* 30 June 1995, pp. 9–11.

USIS, "U.S.-EU Economic Relationship:The NTA Marketplace," *U.S. Information and Texts,* December 7, 1995, pp. 9–10.

USIS, "Joint U.S.-EU Action Plan," *U.S. Information and Texts,* 7 December 1995, pp. 10–19.

Vandenbussche, H., "How Can Japanese and Central European Exporters to the European Union Avoid Antidumping Duties?" *World Competition,* March 1995.

Willig, R.D., *The Economic Effects of Antidumping Policy.* [Mimeo.] OECE, Paris, October 1995.

Yamamura, K., "Caveat Emptor: The Industrial Policy of Japan," in P. Krugman (ed.), *Strategic Trade Policy and the New International Economics.* Cambridge, Mass., 1986.

Zeidler, G., "Distribution Systems and Vertical Integration: Protectionism or 'Normal' Business Practice?" Discussion paper prepared for the conference "Towards a New Global Framework for High-Technology Competition," Kiel, Germany, 30–31 August 1995.